T0341174

Mitigation of Gas Pipeline Integrity Problems

Mitigation of Gas Pipeline Integrity Problems

Mavis Sika Okyere

CRC Press

Taylor & Francis Group

Boca Raton London New York

CRC Press is an imprint of the
Taylor & Francis Group, an **informa** business

First edition published 2021
by CRC Press
6000 Broken Sound Parkway NW, Suite 300, Boca Raton, FL 33487-2742

and by CRC Press
2 Park Square, Milton Park, Abingdon, Oxon, OX14 4RN

© 2021 Taylor & Francis Group, LLC

CRC Press is an imprint of Taylor & Francis Group, LLC

ISBN: 978-0-367-54658-8 (hbk)
ISBN: 978-1-003-09002-1 (ebk)

Typeset in Times
by Deanta Global Publishing Services, Chennai, India

Dedication

Dedicated to my parents with love and gratitude. You are the greatest influence on my academic accomplishments.

The author would also like to dedicate the book to her husband, Yaw, and children, Kwadwo Agyarko and Ama Sarpong, for accompanying her in the journey of writing this book for the pipeline industry professionals.

Contents

Preface

Mitigation activities and solutions minimize the extent or impact of the released volume and related damage. Product releases can have disastrous consequences, so ensuring pipeline integrity is important for pipeline operators. Pipeline integrity is not just about preventing incidents, but is a holistic approach to the prevention, detection, and mitigation of product releases.

The purpose of this book is to present the methodology that will enable an engineer, experienced or not, to alleviate pipeline integrity problems during operation. This book explains the principal considerations and establishes a common approach in tackling technical challenges that may arise during oil and gas production. In view of this, it can definitely serve the needs of engineering students and professionals in the field of pipeline engineering.

The scope of this book is as follows:

- Third-party damage
- Corrosion
- Construction and materials defect
- Geotechnical hazards
- Stress corrosion cracking
- Off-spec sales gas
- Improper design or material selection
- As-built flaws
- Improper operations
- Leak and break detection
- Natural gas hydrate

It offers the tested concepts of pipeline integrity blended with the results of current ideas and recent research, documented in a scholarly fashion to make it simple to the average reader on the subject. Standard and reference materials have been drawn from the works of engineering theorists and scientists.

Mavis Sika Okyere
Ghana National Gas Company

Acknowledgments

I am grateful to God for the good health, wisdom and wellbeing that were needed to write this book.

I want to express my deepest gratitude to Dr. R. Winston Revie, from whom I have gained expert guidance, critical comment, dedicated support, and personal inspiration. I appreciate all those who offered comments, those who assisted in editing and proofreading, and so many people whose names may not all be enumerated. Their advice and contributions are appreciated and gratefully acknowledged.

Thanks to my family and friends who, in different ways, have supported and encouraged me to explore my potential and pursue my dreams.

To my children, Kwadwo Agyarko Nyarko Okyere and Ama Sarpong Okyere; and my husband, Yaw Okyere, you have been my rock and my cheerleader; thank you for your steadfast support.

May God bless all of you!

Author Biography

Mavis Sika Okyere (née Nyarko) is a senior pipeline integrity engineer at Ghana National Gas Company. She is an expert in risk-based assessment, pipeline integrity, corrosion protection, and monitoring. She has experience with subsea structural engineering, piping, and pipeline engineering principles as applied to both onshore and offshore conditions.

Mavis studied MSc gas engineering and management at the University of Salford, United Kingdom, and BSc civil engineering at Kwame Nkrumah University of Science and Technology, Ghana.

She worked with LUDA Development Ltd, Bluecrest College, INTECSEA/WorleyParsons Atlantic Ltd, Technip, Ussuya Ghana Ltd, and Ghana Highway Authority. She has published several books and journals and is a member of many national and international bodies.

1 Introduction

Pipeline integrity does not only concern how to keep the contents inside the pipeline, but also the means to prevent them from escaping. The causes of the loss of containment, which is more commonly described as leaks, have been classified historically into three main areas (other causes of incidents rarely affect the integrity of the pipeline). These are in order of magnitude for the United States/Europe:

- Third-party damage (35–40%)
- Corrosion (20%)
- Material and construction defects (15–20%)
- Others (20%)

It is of interest to note that these percentages change between locations; i.e. in the Union of Soviet Socialist Republics (USSR), due to the large lengths of high-pressure gas pipelines located in sparsely populated and remote regions and historical poor quality control, material and construction defects account for the highest percentage of leaks and ruptures (Posakony 1993).

Third-party damage not only has historically been the most common loss of integrity, but also provides the largest loss of integrity and the highest level of damage. The root cause of many incidents is the lack of knowledge of a pipeline's existence or depth. The most obvious place to start with in terms of information is visual markers on the ground. However, operating companies often find themselves in a dilemma as to whether increased knowledge of the location of the pipelines makes them vulnerable to deliberate attack and whether it outweighs the decrease in third-party activity. Marker posts also require constant maintenance to keep up to date telephone numbers, etc., and to repair damage. Marker tape below ground has been shown to be of little effect unless combined with some other form of protection.

Regular contact and notification with the landowner and the occupier usually reap large dividends in preventing damage. Many pipeline operators have a right to walk and inspect the pipeline annually, which affords them time to contact each landowner. Loss of this right through inactivity and not keeping up to date a list of owners and occupiers have often been shown to be a false economy for the pipeline operator.

A further means of preventing third-party damage which has received much attention in the United States is the use of a "one call" system whereby a contractor can phone one (toll-free) number, giving the location of where he is planning to excavate, and a central record provides details on any pipeline or cable buried in the vicinity. This normally demands legislation to require all operators to provide the information and contribute toward the running costs.

The majority of pipeline companies utilize aerial observation of their pipelines, which can provide warnings of works being undertaken on or close to the pipeline. The normal frequency adopted is bi-weekly, although this can have the effect of third-party activities being kept on hold until helicopter or aeroplane has inspected the right-of-way with the knowledge that there is two weeks' grace before another inspection.

As in most activities, the hardest information to gather is near misses or in the case of pipelines dents or gouges. There is a human tendency not to report damage to someone else's property if it is not apparent and to forget about it. Pipelines are vulnerable to this type of damage, especially gouges that concentrate stress levels and accelerate corrosion at the locality. The only practical way to get information about these types of incidents is for the pipeline operator to accept that repair cost will not fall onto the person inflicting the damage. In the long run this will produce effective results, but is sometimes difficult to accept.

There are few pipelines that are buried without some form of corrosion protection incorporated on them in the form of coating(s) and cathodic protection. The various types of coatings and their advantages are covered in depth in Chapter 3, but two main points can be highlighted. The first is that the long-term success of a coating has been shown many times to be related to the surface preparation. Blast cleaning to ISO 8501-1 (SA 1.0–3.0) standard is a commonly used specification with pictorial and written guidelines. The second is that a cathodic protection system needs regular inspection and monitoring to ensure its continued success. Too low a voltage does not provide protection and too high a voltage can damage coatings quite severely. Internal corrosion can be prevented by the use of the correct material to resist attack, internal coating, or additives to prevent corrosion. Internal coatings have not had a good track record in the past, but advances in application technologies now mean that they should be at least as good as external coatings now are. Corrosion inhibitors have a good track record when used correctly, but require constant injection, albeit of concentrations in the order of 10 parts per million. The inclusion of water within oil and product pipelines can cause considerable problems when a pipeline loaded fully or partially with products is left dormant for long periods of time. The only satisfactory way to mothball a pipeline is to clear the contents and replace them with an inert dry liquid or gas. This is, however, often not feasible for short durations, which over a period of a few years can lead to substantial internal corrosion, commonly located about the 6 o'clock position.

The only way to reduce material and construction defects is with effective quality control.

The ensuing chapters aim to give some guidance on the means of pipeline integrity and monitoring of a pipeline commonly used. It focuses on the mitigation of the challenges to a pipeline's integrity during design, construction, and operation (see Figure 1.1).

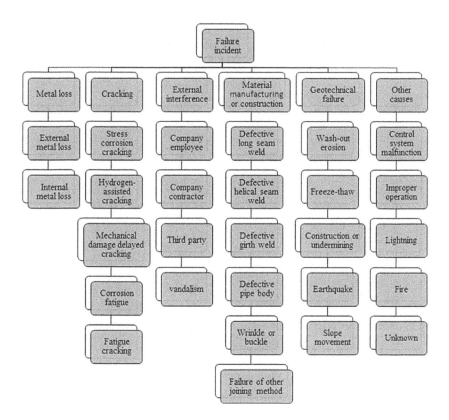

FIGURE 1.1   Challenges to a pipeline's integrity.

2 Third-Party Damage

2.1 INTRODUCTION

Pipeline integrity is an important requirement for the safe and reliable operation of a gas transmission system. Major threats to pipeline integrity include third-party damage, corrosion, stress corrosion cracking, material and construction defects, and soil movements. Of these, third-party damage leading to leaks and ruptures in pipelines, environmental damage, economic loss, and even death is a major threat for buried oil/gas transmission pipeline systems (John et al. 2001).

As shown in Figure 2.1, *third-party damage* may be caused by the following activities on or near the pipeline right-of-way (ROW):

- Operating vehicles or mobile equipment over the right-of-way where a roadway does not exist
- Reducing the depth of soil covering the pipeline
- Plowing deeper than 30 cm (1 foot)
- Ground leveling
- Mining
- Installing drainage systems
- Augering
- Fencing/landscaping
- Ditch digging/clearing/excavation
- Fire

Third-party damage to pipelines can:

- Cause penetration of the pipe wall
- Create dents, gouges, and cracks
- Rupture the pipeline
- Reduce the pressure strength of the pipeline below the maximum allowable operating pressure

This chapter proposes solutions to help pipeline owners and operators to end third-party damage to their pipelines.

2.2 MITIGATION

If through risk assessments or by direct examination you determine the pipeline is vulnerable to third-party damage, one or more of the following mitigation measures should be introduced before or throughout construction and for the duration of pipeline operation (OECD 1997).

1. Public awareness
2. Route selection: avoid populated areas
3. Burial depth: increased depth of burial
4. Provision of additional protection
5. Pipeline surveillance
6. Improvement of signage
7. Right-of-way intrusion detection
8. Notification system
9. Safety management
10. Improvement of material quality
11. Decreased stress
12. Design factor: use of lower design factor

These mitigation measures are described in Sections 2.2.1–2.2.10.

FIGURE 2.1   Pipelines at Tema torched by suspected arsonists (Peacefm 2019).

2.2.1   Public Awareness

Before the commencement of pipeline construction, it is a good practice to consult with local authorities, statutory bodies, and non-statutory groups, and have at least 85% of landowners, pressure groups, and service industries along the pipeline route indicate their willingness to have the pipeline within their land (i.e. approval in principle). However, approval in principle has no bases in law and does not contain any rights to construct the pipeline. These groups should be contacted and given the opportunity to comment on the proposed project at an early stage.

With reference to Table 2.1 of API RP 1162, some examples of public education activities are:

- Producing handouts and brochures, e.g. safety brochures
- Attending exhibitions and presentations
- Issuing environmental leaflets and information
- Liaison with local authorities
- Liaison with landowner/occupiers
- Answering general inquiries from the public
- Advertisements
- Safety videos

If powers are available for the pipeline owner to buy land or acquire rights over the land in which to place a pipeline without the agreement of the landowner or occupier, this should be investigated and used where necessary.

Public awareness programs can be conducted monthly to create awareness among people. Standards, codes, and regulatory policies such as API RP 1162, 49 CFR 195.440, and 49 CFR 192.616 provide guidance for the development and implementation of enhanced public awareness programs.

2.2.2 ROUTING

Pipeline routes are designed to include deviations from the straight line for a variety of reasons:

- Capital costs are generally minimized by the use of "spread" construction techniques, which require unobstructed terrain to allow fast installation rates.
- Public safety requires that minimum proximity distances between the pipeline and occupied buildings be observed.
- Pipeline safety requires that the pipeline be continuously supported along its length and located away from excessive threats of external interference.
- Particular environmental sites or landowner requirements may dictate special methods of working that both increase cost and increase the risk of accidents.

Pipeline routing shares a broad objective with the majority of new developments. The aim is to locate the lowest cost option meeting the safety requirements of both the general public and the facility itself while minimizing the impact on the environment and on owners, occupiers, and authorities who will be required to provide accommodation.

The straight-line route is likely to meet the objective only in the most simplistic and shortest cases. Deviations from the straight-line route are usually required to provide an alignment that is consistent with the overall objective. Such deviations must, however, be justified and defendable, e.g. to a public inquiry. The main constraints to routing can be summarized as follows:

- Codes and legislation
- Public safety
- Pipeline safety
- Environmental considerations
- Landowner/occupier requirements

2.2.2.1   Route Selection

The routing of onshore pipelines is much more flexible than those offshore, but also has to take into account a large number of factors not seen offshore, mainly relating to the occupation of the land by people and existing structures such as roads and buildings. This section is written in general terms for oil and gas pipelines worldwide, but draws on the experience of pipelines in the UK and Europe.

Pipeline routing is inevitably a compromise between opposing factors of minimum route length, avoidance of populated areas, wet or rocky ground or mountainous regions, reduction in major crossings, statutory requirements, and obtaining permission of landowners.

General requirements and identification of parameters to consider when routing pipelines onshore are discussed here. These parameters are numerous, and each pipeline has to be assessed individually. It is difficult to provide examples due to the different parameters to be used and the relative weight given to each of them. The parameters include:

- Legislative requirements
- Landownership and compensation
- Environmental sensitivity
- Habitation
- Land use and type
- Topography
- Crossings
- Safety codes and requirements
- Construction access
- Existing pipelines, utilities, and overhead power cables

Successful pipeline routing is key to providing a pipeline system that optimizes material cost, construction, and safety. For new pipelines, take into consideration the potential for population density increase, thereby avoiding such future problems. Route pipelines away from densely populated areas (e.g. towns, cities, villages) by a sufficient margin to avoid any impact in the future for expansion.

Land use planning considerations should be taken into account both in the routing of new pipelines, in order to limit proximity to populated areas to the extent possible, and in decisions concerning proposals to build in the vicinity of existing pipelines (OECD 1997).

Pipeline routing onshore is the key activity, incorporating a multitude of different factors to be considered and a compromise chosen between some, while compliance with others is a statutory requirement. The ability of pipelines onshore to accommodate sharp changes in direction by the use of forged bends or cold bends allows a large degree of flexibility in routing past objects. The following subsections describe the design method for routing onshore pipelines.

2.2.2.1.1 Desk Study

The first main activity to take place is the desk study. This utilizes all the available paper-based knowledge about the area where the pipeline is planned, predominantly topographical and other maps of varying scale. Maps of minimum scale 1:50,000 are required in order to accomplish the aim of the desk study, which is to identify a potential corridor, or corridors, where further study should be concentrated. The following factors are typical of those addressed at the initial stage (HRR 1992).

2.2.2.1.1.1 Shortest Distance The predominant aim of most cross-country high-pressure pipeline designers is to be able to construct the pipeline in the shortest time for the lowest possible cost. Thus, all other items being equal, the shortest route plotted between the end points of the pipeline would be the cheapest. Apart from desert and bush/scrub locations, this is not normally practical; however, deviations from the straight line should be held to a minimum.

2.2.2.1.1.2 Topography Topography and land type are the next major consideration. Pipeline construction is easiest in flat or gently sloping, well-drained soil. Conditions outside of this are feasible for construction, but the severity of the slope and the type of ground need to be considered when choosing alternatives. Mountainous regions, which are often accompanied by hard rock, cause considerable additional difficulties in construction, which is reflected in the cost, which can be up to ten times greater than that for a similar distance on the easiest sections (HRR 1992).

2.2.2.1.1.3 Crossings The number and type of potential crossings often affect the overall routing of a pipeline onshore. In general, crossings of any major road or river should be avoided where possible, or the number of times it crosses the same feature should be reduced to the minimum. Crossings are often time consuming and disruptive to progress and cost a significant lump sum of money, in addition to the regular pipeline cost, normally based on length.

2.2.2.1.1.4 Inhabited Areas The location factor determines the design factor and hence the wall thickness and cost for sections of the line pipe. Avoidance of habitation for high-pressure gas pipelines is advantageous from this aspect and also from the view of overall pipeline safety. Thus, villages, towns, and cities need to be avoided and often by a sufficient margin so as to avoid any impact in the future for expansion. Plans produced by local authorities in the project area are excellent sources of information regarding planned development areas.

2.2.2.1.1.5 Corridor of Interest   Once this aspect of the design is complete, a corridor of interest should be established, which can then be further investigated. Alternative routes should be identified and should not be discarded too early in the routing process as problems on a particular route often appear later in the process. The reasons for choosing the preferred route above any other may need to be justified to outside bodies, so sufficient evidence should be gathered to support the elimination of a route from further consideration. Equally, too many alternatives dilute the design effort and allow excessive effort to be expended on unsuitable routes.

2.2.2.1.2   Site Reconnaissance

Following the desk study, field reconnaissance is required for the pipeline designer to check the accuracy of the maps used for the study; gain an appreciation of the type of countryside the pipeline is running through; and obtain initial information with regard to land use, the severity of any crossings, the type of ground, potential construction techniques to be used, and available access. Often, additional information can be found from local authorities along the route and initial contact made with relevant bodies and information sources. Pictures, videos, and site notes taken during such visits are invaluable for use later in the design process.

Information gathered during initial reconnaissance visits should include:

- Land type and current use
- Details of crossings (roads, rivers, etc.)
- Evidence of future or current construction work
- Location and use of buildings in the pipeline corridor
- Examination of local maps and sources of information

2.2.2.1.3   Legislation and Land Ownership

Following the site reconnaissance survey, any alternative pipeline corridors should be able to be graded and some eliminated from further effort. The additional information found from the visit should be included in the existing maps or larger-scale maps (1:10,000) and should be used from this point on.

The next logical step in the design is contact with the relevant government department(s), if this has not already occurred. Legislative requirements for permission to construct a pipeline vary considerably between countries, and even within countries. If the pipeline is being constructed by a private company as opposed to a utility company or government agency, the legislation often provides little or no power to impose a route on objecting individuals or authorities. Documentary requirements vary considerably between countries, and local information should be sought to enable the required documents and plans to be submitted in a timely manner.

The position in the UK for gas pipelines is that pipelines laid for a public gas supplier can be laid under the Gas Act, which requires compulsory purchase of land and allows little room for opposition by one or two interested parties to stop the pipeline from being constructed. Other pipelines are built under the 1962 Pipelines Act, which involves an application to the government for pipelines over 10 miles (16 km) in length and to the local authority for planning permission for shorter pipelines (HRR 1992). The act allows compulsory purchase as a final action, but in general allows objections to be sustained by any party and resolution of objections to be undertaken by the pipeline constructor. This can often be a lengthy procedure and may lead to a public inquiry or set of hearings for a pipeline where agreement cannot be reached with either the local planning authority or individual land owners.

In conjunction with the approaches to the authorities, the position with land ownership and rights of occupiers needs to be established for the country where the pipeline is routed. As with the legislative requirements, this varies from country to country and can range from the edict that all land is owned by the government to the situation where multiple owners have small strips of land handed down from generation to generation. The requirements or powers allowed by the legislation have a large effect on the ability of the pipeline designer to route the pipeline in the most effective manner at the least cost. If land prices are not controlled or a general payment level is not accepted by all concerned, obstructive landowners can cause extensive delay or additional cost as the pipeline routing progresses. If powers are available for the pipeline owner to buy land or acquire rights over the land in which to place a pipeline without the agreement of the landowner or occupier, this should be investigated and used where necessary.

2.2.2.1.4   Detail Routing

The next stage of the routing process is to obtain more accurate survey information or aerial photographs of the route corridor on which to mark the route and present information to contractors and other authorities, which act as a base for as-built information and operational maps. These maps are normally at a scale of 1:2,500 or 1:5,000 for general routes and 1:500 or less for crossings or congested areas.

Geotechnical and possibly hydrological information is required in order to provide sufficient information to design crossings, analyze potential soil movements (seismic), allow for potential dewatering, and provide information for the construction contractor. The next stages presented are incorporated into routing decisions at this stage.

Pipeline route drawings, sometimes called strip maps or alignment sheets, provide detailed information about a particular route, giving information on chainage (length from the start), land owner/occupier, land use, reference numbers of crossings, depth of cover, pipeline wall thickness, and special requirements, as a minimum. This information is usually laid out with the map or photo on the top half of a drawing sheet, with the other information defined along a longitudinal strip on the lower half of the drawing.

2.2.2.1.5   Environmental Concerns

Disruption and damage, both short term and long term, to the natural environment are becoming a very important concern for pipeline routing. The requirements vary from country to country, but in many industrialized countries a formal investigation into the potential effects on flora and fauna of construction and operation of a pipeline along the route needs to be carried out. In most instances, pipelines can be shown to cause very little long-term harm, certainly when compared to alternative modes of transport.

Many construction techniques now exist to cross particularly sensitive land, including directional drilling, uplift and removal of pasture land, and more sensitive ditch and stream crossings. From a routing viewpoint, sensitive areas are generally to be avoided, but if unavoidable, the length is to be reduced to a minimum. This includes areas set aside as national parks and other similar designations as these areas often have additional legislation protecting them from development or construction.

An accepted fashion with which to allow routing of a pipeline in sensitive environmental areas is an agreement with concerned environmentalists the means and manner of the reinstatement works following construction. This often requires special areas and means of working and will need to be identified during detail routing.

2.2.2.1.6   Safety Codes and Requirements

The design codes make mention of certain variances being possible, provided that a safety authority can be satisfied by a safety analysis that the variance is acceptable in terms of Health and Safety to the general public. The safety authority may require a safety analysis as a standard to demonstrate the level of risk arising from the construction of a gas pipeline. Particular additional requirements and regulations are dependent on the individual country and can be onerous in terms of items such as maximum pressure, the number of block valve stations, and the application of design codes.

As an example, minimum proximity distances of the pipeline from certain dwellings are required by certain design codes, in particular BS 8010, as a recognition of the potential effect of a gas pipeline rupture for whatever reason. The minimum proximity distances rise according to maximum operating pressure (MOP), pipeline size, and wall thickness. For a large-diameter high-pressure gas pipeline, this can be a considerable distance. Using BS 8010, a 36″ gas pipeline at 100 barg MOP and using a design factor of 0.72 has to be separated by 100 m from any dwelling or parallel to any main roads, and for a 40″ 150 barg gas pipeline this rises to 150 m. The requirements for liquid pipelines tend not to be so restrictive, generally a minimum of 3 m, unless specifically mentioned in the requirements of individual countries.

Safety of the general population and the personnel involved in construction is of primary importance and should always be considered and discussed during the detailed design stage to ensure that the final design is as safe to construct and operate as is reasonably possible.

2.2.2.1.7   Working Width

The pipeline route comprises a strip of land, usually termed the working width. This width allows room for topsoil stripping, the pipeline to be laid prior to construction, subsoil or rock to be stored when the trench is dug, and access along the pipeline route for other construction vehicles. The width of this strip varies as required and is usually extended at major road and river crossings to allow for the additional machinery and required storage areas. For a 36″ pipeline, a working width between 30 and 40 m is common. An allowance for this width needs to be made on the pipeline route drawings.

Access for construction equipment will be required at various points along the pipeline route, usually where the pipeline crosses a suitable road. Where this is not feasible the pipeline route must be capable of supporting access along the route as well as the construction activities going on within the working width and possible inclusion of temporary access roads within the design.

2.2.2.1.8 Existing Pipelines and Utilities

It is common to find other utilities and pipelines either parallel to or crossing the proposed route of the pipeline. In some countries, planning laws dictate the emergence of corridors for all such utilities and pipelines. Close co-operation is required from the utility or other pipeline owners in respect of construction works, available space, and any interference with the cathodic protection (CP) system.

In the case of overhead power lines, it is generally not recommended that pipelines follow them in parallel for any particular distance due to the potential problems associated with induced voltages and current. This, combined with the potential for earth leakage, can cause considerable difficulties for the CP system and can also cause problems with welding and the potential generation of sparks during maintenance operations if the pipeline is cut at any point in the future.

2.2.2.1.9 Route Selection Requirements for an Offshore Pipeline

It should be proposed that the following criteria (where applicable) will be considered during pipeline route selections:

- Route length
- Review of existing survey data
- Existing pipelines, cables, platforms, structures, obstacles, etc.
- Future developments
- Anchoring and/or restricted areas
- Pipeline crossings with third parties
- Potential for self-burial
- Riser locations/topsides tie-ins
- Platform approaches
- Expansion loop dimensions
- Installation aspects
- Protection of the system
- License requirements from regulatory bodies or third parties
- Fishing patterns

The following points may also be required:

- Allowable bend radii
- Pipeline route clearances – minimum clearance for all existing or future developments for the prevention of potential construction vessel anchoring problems
- Pipeline crossings
- Sea bottom variations

2.2.3 BURIAL DEPTH: INCREASED DEPTH OF BURIAL

Onshore pipelines are normally buried for the following reasons:

- Safety
- Protection against damage
- Reduced environmental and aesthetic impact
- Reduced area of land "sterilized" by the pipeline (in terms of limits on land use for buildings, etc.).

The depth to which a pipeline is buried depends on a number of factors and is referenced in the design codes, as shown in Table 2.1. In built-up areas (towns, cities) and close to areas of activity (main roads, etc.), historical evidence has shown that increased cover over the pipeline up to a maximum of 2 m gives significant benefits in reducing the number of impacts from *third parties*, most of which limit their actions to within 1–1.5 m depth. In an evaluation of damage data from most pipeline transmission systems, the likelihood of damage has been shown to reduce by a factor of 10 by increasing the cover from 1.2 to 2.2 m.

FIGURE 2.2   Natural gas pipeline being buried (Source: www.overpipe.com).

As the increased cover has the effect of increasing cost as the trench and amount of the earth to be moved increases proportionally with depth, increase the depth to which the pipeline is buried. For example, for a 36″ pipeline, the trench could increase up to 3.5 m below ground level if a cover of 2.2 m was thought necessary in certain areas. This in itself poses additional dangers to the construction personnel in terms of trench collapse and additional problems in terms of water collection within the trench (Figure 2.2).

TABLE 2.1
Minimum Cover for Buried Pipelines (IGEM 2008, ASME 2010, GSA 2012)

Location	IGE/TD/1 Edition 1	IGE/TD/1 Editions 2, 3 & 4	IGEM/TD/1 Edition 5	PD 8010-1:2015	ASME B31.8	AS 2885.1	L.I. 2189 Normal soil	L.I. 2189 Consolidated rock
All	0.9 m (3 ft)	1.1 m						
Rural areas	–	–	1.1 m	0.9 m	0.61 m (Class 1) 0.76 m (Class 2)	0.75 m	762 mm (Class 1)	457 mm (Class 1)
Suburban areas	–	–	1.1 m	1.2 m	0.76 m (Classes 3 & 4)	0.9 m	914 mm (Classes 2, 3, & 4)	610.9 mm (Classes 2, 3, & 4)
Roads	–	–	1.2 m	1.2 m	0.91 m	–		
Water courses, canals, rivers	–	–	1.2 m	1.2 m		1.2 m	914 mm	610 mm
Railways	–	–	1.4 m	1.4–1.8 m	0.91 m	–		
Rocky ground	–	–	–	0.5 m	–	0.9 m (W) 0.6 m (T1, T2) 0.45 m (R1, R2)		

2.2.4 ADDITIONAL PROTECTION

Generally, provide additional protection to the pipeline by increased wall thickness, concrete slabs, tiles, plates (e.g. overpipe high-density polyethylene plate), high tensile netting, sleeving, etc., in areas considered to have an increased potential for third-party activity (see Figures 2.3 and 2.4).

Reinforced concrete slabs, tiles, or steel plates are buried above the pipeline so that in the event of excavation the slab, tile, or plate is encountered before the pipeline. In theory, damage is limited to the slab, tile, or plate and the pipeline is not affected (John et al. 2001).

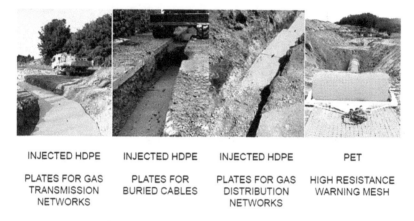

INJECTED HDPE	INJECTED HDPE	INJECTED HDPE	PET
PLATES FOR GAS TRANSMISSION NETWORKS	PLATES FOR BURIED CABLES	PLATES FOR GAS DISTRIBUTION NETWORKS	HIGH RESISTANCE WARNING MESH

FIGURE 2.3   Overpipe HDPE (Source: www.overpipe.com).

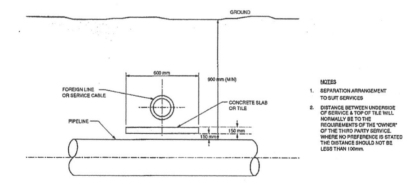

FIGURE 2.4   Typical third-party service crossing (Source: HRR. 1992. *Onshore pipeline design base manual.* Design guideline, JP Kenny).

2.2.5 PIPELINE SURVEILLANCE

Routine surveillance of the complete pipeline should be undertaken throughout its operational life to ensure the continued integrity of the pipeline by the early detection of any loss of cover, subsidence, and any third-party activities on, or in the vicinity of, the pipeline route.

a. Increase the frequency of pipeline right-of-way inspection, e.g. helicopter surveillance, aerial/vehicle/foot patrols, satellite surveillance, and vantage point survey.
   • Line walking: at regular intervals, a representative of the operator should inspect the complete length of the pipeline on foot. To assist in this inspection, the pipelines should be delineated by pipeline marker posts.
   • Helicopter patrols: at regular intervals, a helicopter inspection of the entire pipeline route should be undertaken and video records made. To assist the helicopter patrol in locating the pipeline, aerial marker post designed with a reflective colored (day glow) surface should be positioned at strategic locations.
   • Patrol problem areas at least every three days and one week for other areas.
   • Ensure unapproved accesses are not used to cross the ROW.
   • Prevent unauthorized activities on the ROW.
b. Enhance ROW visibility through the clearing of vegetation covering ROW markers and signs. Weed around the markers 1 m radius by chemical or manual methods (see Figure 2.5).
   • Trees often hide pipeline markers and the corridor that reminds neighbors and contractors of a pipeline in the area. Keeping the pipeline right-of-way clear reduces the risks of third-party damage and increases the safety of all (NEW 2016).
   • Tree roots can damage the coating and come in contact with the pipe steel. Tree roots carry water and nutrients to the rest of the tree and for that reason are very good conductors of electricity. Risks associated with corrosion leaks and corrosion-related pipeline failures are significantly increased when the pipeline coating is damaged and the tree roots absorb the electric current necessary to stop corrosion.
   • Third parties should not plant closer than 25 ft from any transmission pipeline.
c. Liaison and better conversation with landowners to handle land usage issues: have a close liaison with the population along the route.

FIGURE 2.5 Clearing of vegetation covering ROW markers and signs.

2.2.6 IMPROVE SIGNAGE

Improve signage on onshore pipelines by marking pipeline with marker post, pipeline markers, road-crossing markers, river bottom protections, rail-crossing markers, river-crossing markers, etc. (see Figures 2.5 to 2.7). These interventions are meant to exclude unauthorized activities from the pipeline right-of-way and should be monitored regularly.

The pipeline markers should indicate the pipeline owner, telephone number in case of emergency, the product being transported, pipe size, the direction of flow, and the distance in kilometers from the control station.

At yearly intervals, carry out maintenance of these markers. All paint used during this regular maintenance should not be harmful to livestock (Figures 2.8 and 2.9).

FIGURE 2.6   A marker post for a buried natural gas pipeline.

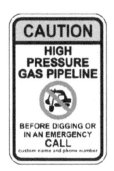  

FIGURE 2.7   Examples of surface markers (Source: www.overpipe.com).

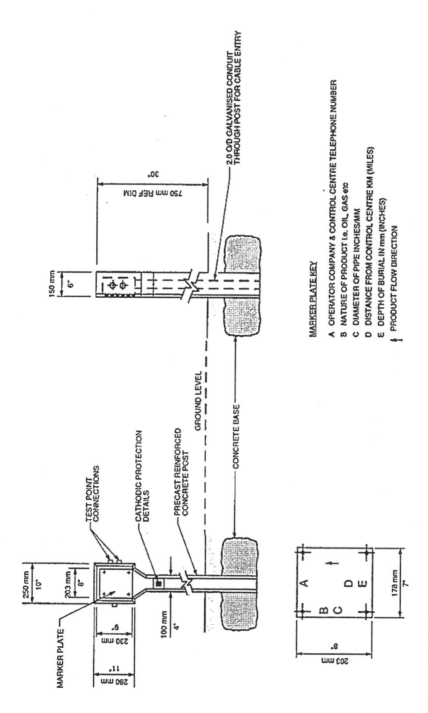

FIGURE 2.8   Pipeline marker and cathodic protection test point.

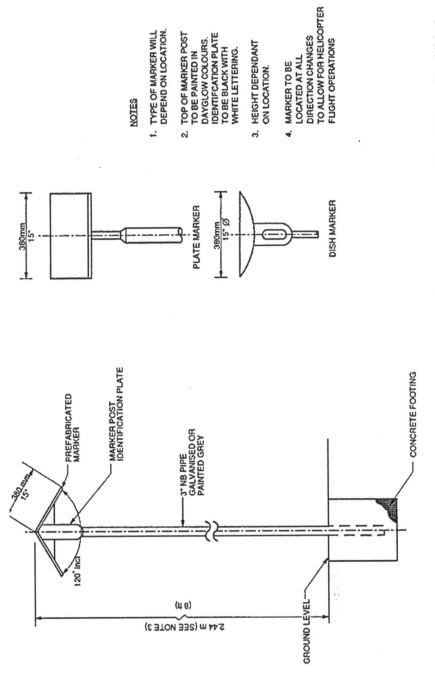

FIGURE 2.9 Typical aerial marker (Source: HRR. 1992. *Onshore pipeline design base manual*. Design guideline, JP Kenny, Unpublished).

2.2.7   RIGHT-OF-WAY INTRUSION DETECTION

Installing an *intrusion detection system/threat scan system* on an aboveground piping system to detect a third-party interference when it happens and precisely locating where it is occurring along the length of a pipeline are vital to mitigating potentially catastrophic situations.

For example, in the case of the Asian gas supplier, Guangdong Natural Gas Group (GDNGG), fiber-optic sensing has proved to be a highly accurate, reliable, and cost-effective tool, which not only can detect leaks with precise location identification but also, when paired with Maxview Integration Software, can alarm personnel to events occurring around the pipeline, proactively preventing damage.

2.2.8   NOTIFICATION SYSTEM

Introduction of a third-party inquiry system, i.e. "call before you dig" or "one-call system" as a law in areas without such legislation. An operator of a pipeline should establish a written program (i.e. Damage Prevention Program) to prevent damage to pipelines and facilities from excavation activities, such as digging, trenching, blasting, boring, tunneling, backfilling, or any other digging activity.

Guide organizations or individuals carrying out works close to pipelines (Lidiak 2010).

In areas that are covered by more than one qualified one-call system, an operator need only join one of the qualified one-call systems if there is a central telephone number for excavators to call for excavation activities, or if the one-call systems in those areas communicate with one another.

2.2.9   SAFETY MANAGEMENT

Design codes of most countries specify what precautions, safety studies, or risk analyses must be undertaken to demonstrate the effectiveness of the various safety measures employed, e.g. minimum distances from buildings. The safety authority may require risk analysis as a standard to demonstrate the level of risk arising from a construction/excavation work along the pipeline. An example of safety management systems used by pipeline operators is the PIMS (Pipeline Integrity Management Systems).

A comprehensive safety analysis/risk analysis is recommended for pipelines that run through or close to populated areas:

- Annual risk assessment of all the threats by a multi-disciplinary team
- Emergency response and control planning reviewed based on risk assessment
- Strengthening administrative controls for condition monitoring
- Focused efforts on surveillance across the pipeline

To manage risk where location class has changed since commissioning:

- De-rate the pipe section to maximum allowable operating pressure (MAOP) applicable to that class location
- Strengthen the pipe, i.e. increase MAOP to the original value
- Increase the existing pipe wall thickness
- Cut and replace with higher-thickness/higher-SMYS pipe
- Re-test the pipe section to establish higher MAOP
- Based on risk assessment, take suitable measures to mitigate risks to acceptable limits. Use integrity assessment methods prescribed by codes and develop and follow a performance plan for risk mitigation

2.2.10   DESIGN FACTOR

Design factor is an important concept in all pipeline codes. Design factor is the ratio between the operating (hoop) stress in the pipeline and its yield stress. Apart from exceptional circumstances, the wall thickness of onshore pipelines is determined by the hoop stress and the choice of design factor, which limits the stress in the pipe to a defined fraction of the specified minimum yield stress (SMYS) of the pipe material. The choice of design factor is limited by the design codes to a maximum of 0.8 or 0.72 and a minimum of 0.4 or 0.3. The lower design factors stated in the codes can be increased in most instances, provided additional measures are taken, or a risk and safety analysis shows that the potential risk is still acceptable when using a higher design factor.

The third-party activity has always been, and remains to this day, the primary cause of major failures in pipelines. It therefore follows that where an increased possibility of third-party activity is found, a larger degree of inbuilt safety against such activity should be imposed. A reduction in the design factor or the use of a lower design factor provides this and is recognized in the codes (OECD 1997) (Table 2.2).

TABLE 2.2
General Design Factors

Location class ANSI B 31.8	Design factor B 31.8 (max)	Location class BS 8010	Design factor BS 8010 (max)
Class 1, Division 1	0.80	Class 1	0.72
Class 1, Division 2	0.72	Class 1	0.72
Class 2	0.60	Class 1	0.72
Class 3	0.50	Class 2	0.30–0.72[a]
Class 4	0.40	Class 3	0.30[b]

[a] Variance to be justified by safety evaluation.

[b] Maximum pressure limited to 7 barg.

The use of a 0.8 design factor in ASME B31.8 is subject to a hydrotest of 1.25 times the design pressure if the MOP is above 72% of the SMYS. Other location classes allow the use of air or gas tests instead of water.

2.3 MITIGATING THIRD-PARTY DAMAGE TO ABOVEGROUND PIPELINES

Aboveground pipelines and components are exposed to threats of vehicular collision and vandalism (see Figure 2.1). In addition to what has been described in Sections 2.2.1 to 2.2.10, the following may be used to mitigate third-party damage to aboveground pipelines and components (Muhlbauer 2004):

1. Barrier-type prevention
   - Electrified fence in proper working condition
   - Strong fence/gate designed to prevent unauthorized entry by humans (e.g. barbed wire, anti-scaling attachments, heavy-gauge wire, thick wood, or other anti-penetration barriers)
   - Normal fencing (chain link, etc.)
   - String locks not easily defeated
   - Guards (professional, competent) or guard dogs (trained)
   - Alarms: deterrent type designed to drive away intruders with lights, sounds, etc.
   - Barriers to prevent forcible entry by vehicles (These may be appropriate in extreme cases. Ditches and other terrain obstacles provide a measure of protection. Barricades that do not allow a direct route into the facility, but instead force a slow, twisting maneuver around the barricades, prevent rapid penetration by a vehicle.)
   - High visibility (difficult to approach the site undetected)
2. Detection-type prevention
   - Staffing: give maximum value for full-time staffing with multiple personnel at all times
   - Video surveillance: real-time monitoring and response, video surveillance for recording purpose only
   - Alarms, with timely response: sound monitors, motion sensors, alarm systems, etc.
   - Supervisory Control and Data Acquisition (SCADA) system: such a system can provide an indication of tampering with equipment because the signal to the control room should change as a transmitter or meter changes
   - Satellite surveillance: with increasingly better resolution, such option is viable today for observing a pipeline and the surrounding area continuously or at any appropriate interval
   - Explosive dye markers: these devices that spray a dye on a perpetrator to facilitate apprehension and prosecution
   - Intrusion detection system

All detection-type prevention must be coupled with timely response.

3. Patrolling: varying the patrol and inspection schedules enhances this as a sabotage preventive measure

4. Simulated measures such as plastic that appear to be steel bars, fake cameras, and signs of warning measures that do not exist: while obviously not as effective as the genuine deterrents, these are still somewhat effective

2.3.1   CASE STUDY: ANOTHER SABOTAGE? GAS PIPELINES AT TEMA TORCHED BY SUSPECTED ARSONISTS

On April 8, 2019, some security men doing some regular checks along the pipelines that run along one of the generating plants within the Tema enclave discovered that the pipelines were burning because some persons had packed car tires over some pipelines that were transmitting fuel to the generation plant and had torched these tires (see Figure 2.10).

FIGURE 2.10   Gas pipelines at Tema torched by suspected arsonists (Peace FM 2019).

2.3.1.1   How Do We Avoid Such Incidents?

Aboveground pipelines and components have different types of third-party damage exposure compared to buried sections. Included in this type of exposure are the threats of vehicular collision, sabotage, and vandalism. Section 2.3 above provides mitigating measures to reduce third-party damage to above ground pipelines.

2.3.1.1.1   Internal Sabotage

An employee with the intent of doing harm is usually in a better position to cause damage due to his likely superior knowledge of the process, equipment, and security obstacles, as well as his unquestioned access to sensitive areas. Some preventive measures are available to the operating company. According to Muhlbauer (2004), common deterrents include:

- Thorough screening of new employees
- Limiting access to the most sensitive areas
- Identification badges
- Training of all employees to be alert to suspicious activities

2.3.1.2   How Do We Fix the Damaged Pipe?

2.3.1.2.1   Cut and Replace

An effective and/or safest way to repair the pipeline defect is to remove the affected segment of pipe (cut the defect) and replace it with a pre-tested section of the sound pipe; tie-in welds should be inspected and returned to service. If the pipe is not pre-tested, then it has to be hydro-tested before the pipeline is returned to normal service. Removal and testing of the pipe section will necessitate the shutdown and depressurization of the pipeline.

Another cost-effective option is to install a sleeve while the pipeline continues operating.

3 Corrosion

3.1 CORROSION MECHANISM

Metal corrosion, from the Latin "corrodere" (to gnaw to pieces), is a deleterious reaction between the metal and its environment. In buried pipelines, it may, if left uncontrolled, cause leaks and thus significant losses of the products transported, plant shutdowns, contamination of products, pollution of soil and groundwater, and even fires. The loss of a few grams of metal may thus result in losses considerably out of proportion to the amount of metal destroyed.

Pipelines may suffer an unacceptable degree of damage from either internal or external corrosion, or both, unless effective measures are taken. It is therefore essential in a corrosive environment to prevent, or limit, the incidence of corrosion by coating and/or other preventative measures.

The effect of corrosion on oil and gas pipelines can be catastrophic. The destruction of the metal eventually leads to leakages, which have the potential to cause massive disasters such as fires and explosions. High safety concerns and strict standards applicable to the oil and gas industry mean that constant monitoring is needed to identify the presence and extent of corrosion.

3.1.1 BASICS OF AQUEOUS METALLIC CORROSION

A corroding system is driven by two spontaneous reactions that take place at the interface between the metal and an aqueous environment. The two simultaneous reactions are the oxidation (anodic) reaction and a reduction (or cathodic) reaction. The first reaction occurs when the chemical species from the aqueous environment removes electrons from the metal; the other is a reaction in which atoms on the metal surface participate to replenish the electron deficiency.

An example of anodic (oxidation, electron-donating) reaction:

$$Fe \rightarrow Fe^{2+} + 2e^- \ (\text{metal dissolution})$$

Examples of cathodic (reduction, electron-accepting) reactions:

$$O_2 + 2H_2O + 4e^- \rightarrow 4OH^- \ (\text{oxygen reduction})$$

$$2H^+ + 2e^- \rightarrow H_2 \ (\text{hydrogen evolution})$$

An important effect is the electron exchange between the two reactions that constitute an electronic current at the metal surface. This, in turn, imposes an electric potential of such a value on the metal surface that the supply and demand for electrons in the two coupled reactions are balanced.

The potential imposed on the metal is of much greater significance than simply to balance the complementary reactions which produce it. This is because it is one of the principal factors determining what the reactions will be.

Generally, at the anodic locations, there is corrosion damage (e.g. metal loss), while at the cathodic location, no corrosion damage occurs. The location of anodes and cathodes tends to move randomly over the surface of the metal for alloys that are subject to general corrosion. However, the location of an anode tends to become strongly localized for corrosion-resistant alloys, which are covered by a passive oxide film. This gives rise to localized corrosion damage such as pitting corrosion, stress corrosion cracking, and crevice corrosion (Figure 3.1).

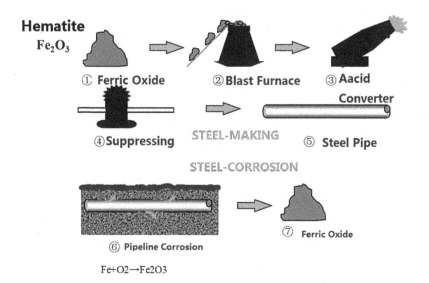

FIGURE 3.1   Basics of corrosion.

3.1.2 FORMS OF CORROSION

Corrosion is generally divided into general corrosion and localized corrosion. Localized corrosion is classified as follows:

   a. Microbiological corrosion
   b. Galvanic corrosion
   c. Crevice corrosion
   d. Pitting corrosion
   e. Intergranular corrosion
   f. Erosion corrosion
   g. Selective corrosion
   h. Stress corrosion cracking (SCC)
   i. Fatigue corrosion

This classification of corrosion is based on visual characteristics of morphology of attack as well as the type of environment to which the metal is exposed.

Microbiological corrosion, also called bacterial corrosion, bio-corrosion, microbiologically influenced corrosion, or microbial-induced corrosion (MIC), is corrosion caused or promoted by microorganisms. It can apply to both metals and non-metallic materials.

Crevice corrosion is a localized attack on a metal adjacent to the crevice between two joining surfaces (two metals or metal–nonmetal crevices). The corrosion is generally confined to one localized area of one metal (Natarajan, 2012). This type of corrosion can be initiated by concentration gradients (due to ions or oxygen). Accumulation of chlorides inside the crevice will aggravate damage. Various factors influence crevice corrosion, such as:

- Materials: alloy composition and metallographic structure
- Environmental conditions such as pH, oxygen concentration, halide concentrations, and temperature
- Geometrical features of crevices and surface roughness
- Metal-to-metal or metal-to-nonmetal type

Filiform corrosion is a special type of crevice corrosion (Figure 3.2).

Pitting corrosion is a localized phenomenon confined to smaller areas. Formation of micro-pits can be very damaging. Pitting factor (the ratio of the deepest pit to the average penetration) can be used to evaluate the severity of pitting corrosion, which is usually observed in passive metals and alloys. Concentration cells involving oxygen gradients or ion gradients can initiate pitting through the generation of anodic and cathodic areas. Chloride ions are damaging to passive films and can make pit formation auto-catalytic. Pitting tendency can be predicted through the measurement of pitting potentials. Similarly, the critical pitting temperature is also a useful parameter.

Galvanic corrosion, often referred to as dissimilar metal corrosion, occurs in galvanic couples where the active one corrodes. EMF series (thermodynamic) and

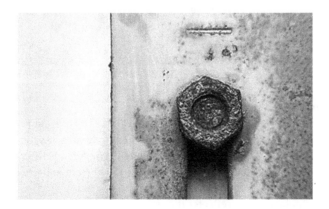

FIGURE 3.2   Crevice corrosion (Source: www.corrosionpedia.com/whats-the-inside-scoop -on-crevice-corrosion/2/6588).

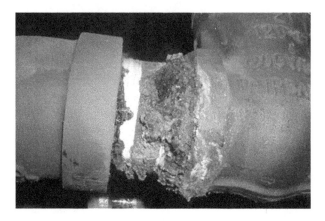

FIGURE 3.3 Galvanic corrosion (Source: www.corrosionpedia.com/21-types-of-pipe-corrosion-failure/2/1484).

galvanic series (kinetic) could be used for the prediction of this type of corrosion (Table 3.1). Galvanic corrosion can occur in multiphase alloys (Natarajan 2012) (Figure 3.3).

Erosion corrosion is the deterioration of metals and alloys due to relative movement between surfaces and corrosive fluids. Depending on the rate of this movement, abrasion takes place. This type of corrosion is characterized by grooves and surface patterns having directionality. Typical examples are:

• Stainless alloy pump impeller
• Condenser tube walls

All equipment types exposed to moving fluids are prone to erosion corrosion. Many failures can be attributed to impingement (impingement attack). Erosion corrosion due to high-velocity impingement occurs in steam condenser tubes, slide valves in petroleum refinery at high temperature, inlet pipes, cyclones, and steam turbine blades. Cavitation damage can be classified as a special form of erosion corrosion.

3.1.3 POLARIZATION AND CORROSION RATES

3.1.3.1 Concept of Polarization

Polarization refers to electrode potential variation caused by current flowing through the electrode, including anodic polarization and cathode polarization. Both anodic polarization and cathode polarization can reduce the potential difference between two electrodes of a corrosion cell, reducing corrosion current and preventing corrosion process.

Essentially, polarization reflects the suppression of the electrode process. Polarization degree reflects resistance degree, while over-potential of the electrode reaction reflects impetus of this electrode reaction.

3.1.3.1.1 Causes of Cathodic Polarization

Cathodic process is a reduction process in which the substance that can absorb electrons in the solution obtains electrons. Causes of its potential becoming negative and polarized are not unitary.

a. Electrochemical polarization: usually, the speed of reduction reaction combining oxidants and electrons at the cathode is slower than the speed of electrons flowing from the anode, resulting in electron accumulation and potential becoming negative, which is called electrochemical polarization.

b. Concentration polarization: concentration polarization happens when hydrogen ions generated at the anode are attracted to the cathode and the hydroxide ions generated at the cathode are attracted to the anode.

3.1.3.1.2 Polarization Diagram

Polarization diagram is used to analyze the main controlling factor of the corrosion process, illustrating the physical meaning of the corrosion potential of a different corrosion system. If the corrosion potential is close to the balance potential of local anode reaction, the corrosion process is determined by cathodic polarization. Meanwhile self-corrosion current is determined by cathode polarization, as shown in Figure 3.4, and the corrosion current is decided by anode polarization, as shown in Figure 3.5.

In corrosion cells, if other conditions are the same, the smaller the polarization of the local cathode or anode is, the bigger the self-corrosion current is.

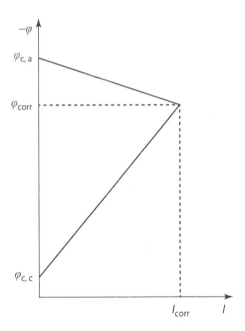

FIGURE 3.4 Cathodic polarization control (Source: Tong, Shan. 2015. *Cathodic protection*. Training document, Ghana: Sinopec).

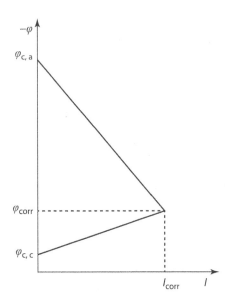

FIGURE 3.5 Anode polarization control (Source: Tong, Shan. 2015. *Cathodic protection.* Training document, Ghana: Sinopec).

3.1.3.2 Corrosion Rate

Because of this process, electric current flows through the interconnection between cathode and anode. The cathodic area is protected from corrosion damage at the expense of the metal, which is consumed at the anode. The amount of metal lost is directly proportional to the current flow. Mild steel is lost at approximately 20 pounds for each ampere flowing for a year (Tong 2015).

The degree of polarization is a measure of how the rates of the anodic and cathodic reactions are retarded by various environmental (concentration of metal ions) and/ or surface process factors. Hence, polarization measurements can thereby determine the rate of the reactions that are involved in the corrosion process – the corrosion rate. The corrosion rate can be expressed in mil per year. Corrosion rate is calculated as

$$\text{Corrosion rate} = \frac{KW}{(\rho At)} \tag{3.1}$$

where
  K = constant which depends on the system used
  W = weight loss after exposure to environment (mg)
  ρ = density of the material (g/cm^3)
  A = area of the material exposed (square meter)
  t = time of exposure (hour)

3.1.3.3 Factors Affecting Corrosion Rate

  a. Potential difference between anode and cathode (galvanic series): Interconnecting two dissimilar metals in an electrolyte will create a corrosion cell. The strength of this cell increases as the distance within the galvanic series increases (see Table 3.1).

b. Circuit resistance – resistivity of the electrolyte

Circuit resistance includes the following:

- Resistance of the anode
- Resistance of the cathode
- Resistance of the electrolyte
- Resistance of the metallic path

Increasing resistance will reduce the corrosion rate.

c. Environmental conditions
d. Stray currents

Example: connecting magnesium to copper will produce a corrosion cell with a potential of about 1.5 V.

TABLE 3.1
Galvanic Series

Metal	Volts (CSE)
Commercially pure magnesium	−1.75
Magnesium alloy	−1.60
Zinc	−1.10
Aluminum alloy	−1.05
Commercially pure aluminum	−0.80
Mild steel (clean and shiny)	−0.50–−0.80
Mild steel (rusted)	−0.20–−0.50
Cast iron (not graphitized)	−0.50
Lead	−0.50
Mild steel in concrete	−0.20
Copper, brass, and bronze	−0.20
High silicon cast iron	−0.20
Carbon, graphite, and coke	+0.30

3.1.4   HYDROGEN – RELEASE CORROSION AND OXYGEN- CONSUMPTION CORROSION

3.1.4.1   Hydrogen – Release Corrosion

Hydrogen-released corrosion, also known as hydrogen depolarization corrosion, is the corrosion of which cathodic process is the reduction reaction of hydrogen ion (Figure 3.6).

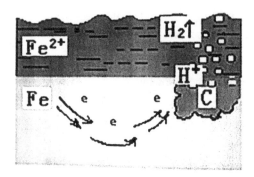

FIGURE 3.6 Hydrogen-release corrosion (Source: Tong, Shan. 2015. *Cathodic protection.* Training document, Ghana: Sinopec).

Electrode reaction:

a. Necessary conditions
   * There must exist H+ ions in electrolyte solution.
   * Anode potential (EA) must be smaller than hydrogen deposition potential (EH), that is EA < EH.
     Hydrogen-releasing potential refers to the potential difference between the balance potential of hydrogen electrode and hydrogen over-potential of cathode under a certain cathodic current density.
b. Features
   * Hydrogen-release corrosion is mainly activation polarization, and concentration polarization can be ignored if there is no deactivation diaphragm on the surface of the metal.
   * Corrosion of metal in acid solution is related to the pH value of solution.
   * The hydrogen-released corrosion of metal in acid solution is generally macro uniform corrosion phenomenon.
c. Main factors that affect hydrogen over-potential
   * Current density
   * Electrode material
   * Electrolyte solution composition and temperature

3.1.4.2   Oxygen-Consumption Corrosion

Oxygen-consumption corrosion is the corrosion of which the cathodic process is an oxygen reduction reaction (Figure 3.7).

Features: compared to the hydrogen ion reduction reaction, oxygen reduction reaction can be implemented at a higher potential, so that oxygen-consumption corrosion is more universal.

a. Necessary conditions
   * There must exist oxygen in electrolyte solution.
   * Anode potential must be lower than oxygen ionization potential, that is EA < EO; oxygen ionization potential EO refers to the potential difference of oxygen balance potential and oxygen ionization over-potential under certain current density.

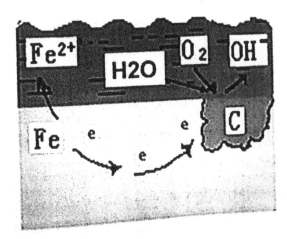

FIGURE 3.7 Oxygen-consumption corrosion (Source: Tong, Shan. 2015. *Cathodic protection*. Training document, Ghana: Sinopec).

TABLE 3.2
Comparison of Oxygen-Consumption and Hydrogen-Release Corrosion

Item	Hydrogen-release corrosion	Oxygen-consumption corrosion
Property of depolarizer	Three ways of mass transfer of hydrogen ion: convection, diffusion, and electro-migration; the diffusion coefficient is very big.	Neutral oxygen molecule, only two ways of mass transfer: convection and diffusion. The Diffusion coefficient is very small.
Concentration of depolarizer	Large concentration; depolarizer is hydrogen ion in acid solution and water molecule in neutral or alkaline solution.	Small concentrations; under ambient temperature and pressure, the saturation concentration of oxygen in neutral water is $10–4$ mol/L, and the solubility of oxygen will decrease as temperature or salt concentration increases.
Cathodic reaction product	Hydrogen gas released in bubble form.	Water molecule or hydroxyl ion; ways of leaving metal surface; convection, diffusion, and electro-migration.
Controlling types of corrosion	Cathodic control or hybrid cathodic control; most is cathodic control, especially cathodic polarization control.	Cathodic control.
Influence of alloying elements or impurities	Obvious influence.	Less influence.
Corrosion rate	If there is no deactivation phenomenon, the corrosion rate of simplex hydrogen-released corrosion is fast because of large hydrogen ion concentration and diffusion coefficient.	If there is no deactivation phenomenon, the corrosion rate of simplex aerobic corrosion is slow because of little oxygen concentration and diffusion coefficient.

b. Features
  - If there is oxygen in the solution, oxygen-consumption corrosion will generate in the first place.
  - As oxygen diffusing is a stable state, oxygen-consumption corrosion is determined by oxygen concentration polarization.
  - Dual function of oxygen: oxygen may play a corrosive role of for easy passivation of metal; it may also play a role of a retardant (Figure 3.7) (Table 3.2).

3.1.5 CAUSES OF CORROSION

When steel is placed in an aqueous environment, an electrochemical reaction takes place on its surface. These consist of cathodic reaction, in which electrons are consumed, and anodic reaction in which the metal is dissolved. The cathodic and anodic reactions are linked, and their rates are related. Therefore, although the anodic reaction is the direct cause of corrosion, the rate of corrosion can be controlled by interfering with either reaction.

The corrosion which results from these electrochemical reactions may appear in various forms, e.g. general corrosion and pitting corrosion. The rate at which corrosion takes place is affected by factors such as the exact composition of the aqueous environment and the temperature. These factors need to be considered when designing corrosion protection systems, although they may not affect the type of protection used.

Other forms of corrosion, while resulting from electrochemical reactions, are more affected by design factors. Examples of these forms of corrosion are crevice corrosion and galvanic corrosion.

Corrosion is an electrochemical process occurring at the interface between metal and environment (Figure 3.8). Three conditions must be present for this to occur.

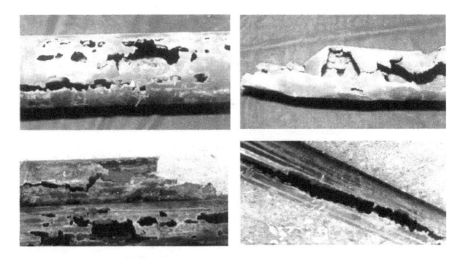

FIGURE 3.8   Pipeline (equipment) corrosion pictures (Source: Tong, Shan. 2015. *Cathodic protection*. Training document, Ghana: Sinopec).

1. Two areas on a structure or two structures must differ in electrical potential (Figure 3.9).

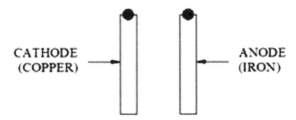

FIGURE 3.9 Two areas or two structures that differ in electrical potential (different amounts of stored energy).

2. Those areas, called *anodes* and *cathodes*, must be electrically interconnected (Figure 3.10).

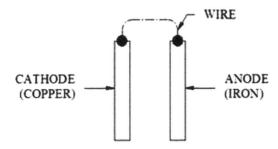

FIGURE 3.10 The two areas, called the "anode" and "cathode", must be electrically connected (conductive path).

3. Those areas must be exposed to a common electrolyte (soil or water) (Figures 3.11 and 3.12).

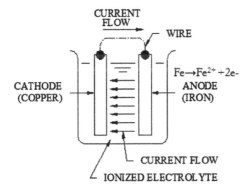

FIGURE 3.11 Areas anode and cathode must be exposed to a common electrolyte.

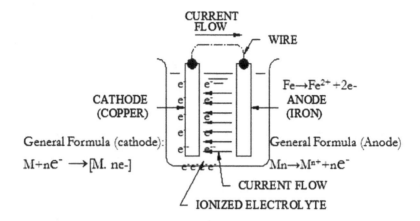

FIGURE 3.12   Corrosion cell – dissimilar metals.

Corrosion cell – dissimilar metals: steel pipe connected to the copper ground rod. Corrosion is also caused by the formation of bacteria with an affinity for metals on the surface of the steel.

3.1.5.1   Corrosive Environments

The deterioration of material due to corrosion can be caused by a wide variety of environments. Metallic corrosion in aqueous (i.e. water-containing) environments occurs with or without dissolved species such as electrolytes (i.e. salts) and reactants (e.g. dissolved oxygen). However, every industry such as the food, pharmaceutical, oil, and gas features a variety of applications encompassing a range of corrosive environments. In the food industry, the corrosion environment often involves moderately to highly concentrated chlorides on the process side, often mixed with significant concentrations of organic acids. The process environment for the pharmaceutical industry can include complex organic compounds, strong acids, and chloride solutions comparable to seawater. The consideration for this work is mostly focused on the oil and gas industry, which has a corrosive environment such as the sour and sweet environment from the sour reservoirs (those containing hydrogen sulfide, H_2S) and the sweet reservoirs (containing carbon dioxide, CO_2). The sour environments can result in sulfide stress cracking of susceptible materials (Taiwo 2013).

3.2   EXTERNAL CORROSION MITIGATION

Effective mitigation of external corrosion involves:

- Selection of appropriate materials
- Use of protective coatings

- Cathodic protection technique
- Adequate corrosion monitoring and inspection

Corrosion of buried pipelines is controlled by the use of protective coatings and maintaining adequate levels of cathodic protection (CP). Materials selection and coatings are the first line of defense against external corrosion. Because perfect coatings are not feasible, cathodic protection must be used in conjunction with coatings. Cathodic protection involves applying a current to the pipeline through the soil from an external source and thus overriding the local anodes, rendering the entire exposed pipeline surface cathodic. Coatings function to separate the steel from the electrolyte and thus prevent corrosion.

The subsequent sections describe the material selection, coating, and cathodic protection as applied to pipelines, subsea equipment, and structures.

3.2.1 MATERIALS SELECTION

Use corrosion-resistant alloys, non-metallic materials (e.g. reinforced composite, thermoplastic, lined, and polyethylene pipelines). Use internally coated carbon steel pipeline systems (e.g. nylon or epoxy coated) with an engineered jointing system.

Material selection and anti-corrosion should be considered in pipeline design. Given that the surrounding of long-distance pipeline is hard to control, surface coating and electrochemistry protection are major projects of pipeline anti-corrosion.

Materials selection is critical to preventing many types of failures. Factors that influence materials selection are:

- Corrosion resistance in the environment
- Availability of design and test data
- Mechanical properties
- Cost
- Availability
- Maintainability
- Compatibility with other system components
- Life expectancy
- Reliability
- Appearance

Choose materials inherently resistant to corrosion in certain environments.

Material selection involves picking an engineering material, either metal alloy or nonmetal that is inherently resistant to the particular corrosive environment and meets other criteria. Variables that will affect corrosion are established along with materials that may provide suitable resistance for those conditions. Obviously, other requirements such as cost and mechanical properties of the potential materials must be considered (Davis 1994).

Data needed to thoroughly define the corrosive environment include many of its chemical and physical characteristics plus application variables such as its velocity. In addition, possible extremes are caused by upset conditions. Non-corrosion considerations include mechanical strength, type of expected loading, and possibly the compatibility of the different candidates with the required fabrication method. After these criteria and other unique ones are considered, the list of materials that can generally satisfy all requirements usually becomes short. Final selection is then made, but trade-off from the optimal meeting of each criterion often is necessary.

Corrosion in oil and gas industry may be prevented by upgrading to more corrosion-resistant materials (stainless steel, nickel alloys), or the corrosive environment may be treated with corrosion inhibitors or oxygen scavengers so that the traditional materials can still be used.

3.2.1.1   Considerations for Material Selection

The pipe material used is selected based on process conditions, the material to be conveyed, and the ability to withstand fire and corrosion. While carbon and stainless steels are commonly used materials of construction, non-metallic and lined or plastic process equipment is increasingly used. The selection of the material of construction should take into account worst-case process conditions that may occur under foreseeable upset conditions and should be applied to all components including valves, pipe fittings, instruments, and gauges. Both composition (e.g. chlorides, moisture) and temperature deviations can have a significantly direct effect on the rate of corrosion. The operator should demonstrate that procedures are in place to ensure that potential deviations in process conditions such as fluid temperature, pressure, and composition are identified by competent persons and assessed in relation to the selection of materials of construction for pipework systems.

A wide range of plastics are available for use as materials of construction and can be used in areas such as handling inorganic salt solutions where metals are unsuitable. The use of plastic linings is widespread in equipment such as tanks, pipes, and drums. However, their use is limited to moderate temperatures, and they are generally unsuitable for use in abrasive duties. Some of the more commonly used plastics are polyvinyl chloride (PVC), PTFE, and polypropylene.

Special glasses can be bonded to steel, providing an impervious liner. Glass or "epoxy" lined equipment is widely used in severely corrosive acid duties. The glass lining can be easily damaged, and careful attention is required.

The following are to be considered for selecting corrosion-resistant materials for construction:

- Type of fluid to be transported (hydrocarbon, water, steam, natural gas)
- Cost-effectiveness
- Flow characteristics (laminar or turbulent flow)
- Temperature
- Pressure
- Properties of material needed
- Application of material
- Product of material (e.g. fittings, valve)
- Design

- Material versus material comparisons to find the best solution
- Industry requirements and regulations

For material to be used at room temperature, the following are the main requirements for materials selection:

- Mechanical properties of the materials: strength, hardness, toughness, resistance to fatigue, and fracture
- Suitable physical properties
- Ease of availability
- Easy formability
- Cost-effectiveness
- Corrosion resistance

When the materials are to be selected for high-temperature components, the requirements for material selection are:

- Mechanical properties: strength (sustaining strength at that temperature)
- Microstructural stability: change of microstructure with temperature is not acceptable
- Creep: the tendency of the material to very slow deformation
- Corrosion, oxidation, and high-temperature corrosion resistance

Materials available for use in the oil and gas industry include:

- Ferrous: steel, stainless steels, superalloys
- Nonferrous: aluminum, copper, titanium, cobalt, and their alloys
- Plastics
- Ceramics
- Composites

The most commonly used material in the oil and gas industry is steel and its various types. Steel is prone to corrosion; therefore all steels used need to be protected by one or more of the methods of corrosion protection: coatings, cathodic protection, inhibitor, etc. (Reza, Chikezie, and Henry 2013).

It is important to select materials based on the requirements of the following international standards:

- National Association For Corrosion Engineers (NACE)
- NACE MR 175
- American Petroleum Institute (API)
- American Society for Testing and Materials (ASTM)
- American Welding Society (AWS)
- Manufacturers Standardization Society (MSS)
- American Water Works Association
- Society of Automotive Engineers (SAE)
- ASME B31

3.2.1.1.1 Material Selection Criteria for Metal Alloys

Steels are used commonly for most oil and gas applications such as piles, pipelines, piping, superstructures, support structures, and storage facilities. The main material selection requirements for steel are:

- Cost-effectiveness
- Mechanical properties such as hardness, percentage elongation, yield strength, tensile strength, density, fracture toughness, creep, fatigue, and brittleness
- Thermal properties such as thermal conductivity, heat capacity, and thermal diffusivity
- Electromagnetic properties such as magnetic permeability/hysteresis, dielectric strength, transparency/color, and electrical conductivity
- Resistance to both internal and external corrosion due to the environment (e.g. soil, groundwater, seawater, salt air)
- Ease of availability
- Fabrication
- Service performance
- Environment: operating temperature range, chemical resistance, radiation resistance, appearance

Steel may be classified as mild-, medium-, or high-carbon steel, based on the percentage of carbon they contain (Reza, Chikezie, and Henry 2013).

Low carbon steel has been identified as having a particular susceptibility to oxidation in the presence of electrolytes, water, and carbon dioxide. Therefore, the problems caused by corrosion mean that an effective solution is imperative to the prevention of accidents resulting from leakages and fractures.

3.2.1.2 Materials with High Corrosion Resistance

Metals that hold high-corrosion-resistant properties include Incoloy 825, Inconel 625, Hastelloy C-276, titanium, super duplex, duplex, stainless steel, and 6MO.

The most commonly specified material for use in the manufacture of corrosion-resistant high-pressure valves and fittings is 316 stainless steel. It is easy to machine, cast, and forge and is readily available as a raw material in a wide range of shapes and sizes.

6MO is 6% molybdenum-alloyed austenitic stainless steel. The increased molybdenum content, up to 6% from the 2% as found in 316 stainless steel, plus the increase in chromium and added nitrogen, provides 6MO with some worthwhile benefits over 316 stainless. Its corrosion resistance is increased particularly with regard to pitting and crevice corrosion. The strength of 6MO is also slightly increased above that of 316 stainless.

Monel alloy 400 is still an excellent choice where atmospheric corrosion is an issue, particularly seawater.

Titanium: Upon exposure to air, titanium immediately forms a passive oxide coating that protects the bulk metal from further oxidation. This protective layer renders titanium immune from attack by most acids and chlorine solutions. Since titanium depends on the presence of an oxide film for its corrosion resistance, it follows that it is significantly more resistant to attack in oxidizing solutions.

Duplex and super duplex: These materials are a natural step up from normal stainless steels and 6MO stainless. Their structure is an amalgamation in equal parts of austenite and ferrite. As a result these high chromium alloys have a higher tensile strength and good corrosion resistance. The super duplex alloy has excellent atmospheric corrosion resistance and finds many uses in aggressive and seawater environments. It is particularly beneficial in resisting stress corrosion cracking, which is a common problem with austenitic stainless steels. Although generally specified for use in structures and fabrication work, it is often used for pipeline components.

Incoloy 825: this alloy will give excellent corrosion resistance to a broad base of severe environments. It has good resistance to stress corrosion cracking, and the high nickel content in conjunction with the copper and molybdenum gives it outstanding resistance to reducing environments such as those containing sulfuric and phosphoric acids.

Stainless steel: Steel alloyed with at least 10% chromium is called stainless steel. The addition of chromium causes the formation of a stable, very thin (few nanometers) oxide layer (passivation layer) on the surface. Stainless steel therefore does not readily corrode or stain when in contacts with water like carbon steel does. But, under some circumstances, the passivation layer can break down, causing local attacks such as pitting corrosion. Pitting corrosion, the predominant form of corrosion of stainless steel, does not allow lifetime prediction as is possible with zinc coatings. The resistance of stainless steel against pitting corrosion can be roughly estimated by the PREN (pitting resistance equivalent number). The PREN is based on the chemical composition of steel, taking into account the amount of chromium, molybdenum, and nitrogen. In literature, various equations for this calculation are given. The most common equations are:

For stainless steels with Mo < 3%

$$PREN = \%Cr + 3.3 \times \%Mo \qquad (3.2)$$

For stainless steels with Mo ≥ 3%

$$PREN = \%Cr + 3.3 \times \%Mo + 30 \times \%N \qquad (3.3)$$

There are various grades of stainless steel with different levels of stability. The most common grade is alloyed with around 18% Cr and 10% Ni. Increasing or reducing the amount of specific elements in the steel changes its corrosion properties, its mechanical properties, or some processing properties such as weldability (Hilti 2015). If the nickel content is significantly reduced, the alloy phase will no longer be purely austenitic but will then combine austenitic and ferritic phases (duplex stainless steel).

Refer to the EN ISO 3506-1:2009 standard (mechanical properties of corrosion-resistant stainless steel fasteners, bolts, screws, and studs).

3.2.2 PROTECTIVE COATINGS

A coating provides a physical barrier between the steel and the aqueous environment and interferes with the flow of electrons.

This section deals with coating materials used for onshore and offshore pipelines and gives information on the various coating systems available. Coating application, repair, use with thermal insulation coatings, underconcrete weight coatings, and in conjunction with cathodic protection are discussed.

Coatings are applied to pipelines in order to insulate the steel surface from the corrosive environment, thereby preventing corrosion and subsequent failure of the structure. Before choosing a coating material to be applied to pipelines and structures, it is important to consider the following factors

1. The environment surrounding the pipe
2. Pipeline-operating temperature (product temperature)
3. Ease of application (yard or site): the coating should be capable of application in the factory or the field at a reasonable rate, and it should be possible to handle the pipe as soon as possible without damaging the coating
4. Inspection methods and ease of repair of the defect and pipe joints
5. Comparative coating system costs
6. Coating properties
   • The coating must be resistant to the action of soil bacteria.
   • Water resistance: the coating must demonstrate negligible absorption of water and must be highly impermeable to water or water vapor transmission.
   • Resistance to disbonding: the coating must be capable of withstanding impacts without cracking or disbonding.
   • The coating must be capable of withstanding stresses imposed on it by soil/pipe movement.
   • The coating must be able to form an excellent bond to the pipe steel with the use of a suitable primer where necessary. In addition, for multi-layer systems, there must be good adhesion between the different coating layers.
   • The coating should show no tendency to creep under prevailing environmental conditions and must have sufficient resistance not to be displaced from the underside of large-diameter pipes.
   • The coating must have sufficient roughness to prevent slippage between the concrete coating and the anti-corrosion coating.
   • The coating must be an electrical insulator of high dielectric strength and must not contain any conducting materials.
   • The coating must be sufficiently flexible to withstand and deformation due to bending, hydrotesting, laying, or any expansion and contraction due to temperature change; cracks must not occur during cooling after application or curing.
   • The properties and adhesion qualities of the coating must be sufficient to withstand certain methods of installation, particularly where pipelay is by bottom tow.

- Resistance to flow: the coating should show no tendency to flow from the pipe under the prevailing environmental conditions, and must have sufficient resistance not to be displaced from the underside of large-diameter pipes.
7. Environmental and health constraints.
8. Pipeline design life.
9. Ease of field jointing.
10. Conformance to standards such as DIN 30670 and NFA 49-711.

The material used for coating pipes varied over the years as technology evolved. For example, in the 1940s and 1950s coal tar, wax, and vinyl tape were used; in the 1960s asphalts were used; and in the 1970s to present day fusion bond epoxy was and is being used. Polyethylene tape and extruded polyethylene jacket material were also used from the early 1950s to the present day (Ginzel and Kanters 2002). Coating technology has improved over time, which has resulted in the use of different types of coating systems, including but not limited to the ones discussed in the following sections.

3.2.2.1 Coal Tar and Asphalt Coatings

These hot-applied enamel coatings have a long and successful record of accomplishment throughout the world, both onshore and offshore. However, the recognition of possible health hazards during the application of asphalt enamel has contributed to the reduction/withdrawal of their use. The Standards BS 4147 (BS EN 10300:2005) is relevant to the materials and applications of these coatings.

- Coal tar enamels generally absorb less water as compared to asphalts.
- Coal tar enamels adhere better to clean steel than asphaltic enamels although this difference is not marked under normal pipeline conditions.
- Although both enamels are soluble in chemical solvents, asphalt is dissolved by petroleum products while coal tar is only softened.
- These coatings should not be used where the pipe operating temperature is likely to exceed 60°C for normal grade coatings. However, higher temperatures can be withstood with higher-grade coatings.
- Both coal tar enamel and asphalt coatings are still widely used for submarine lines under concrete weight coatings.

3.2.2.1.1 Coal Tar Enamel

Coal tar is a byproduct of coke production. The coal tar pitch is mixed with coal tar oils and inert filler such as talc or slate dust to produce coal tar enamel. The exact quantities affect properties such as hardness and flexibility.

After cleaning the pipe, a quick drying primer is applied and allowed to dry. Following this, the coal tar is applied by flooding, and during this process layers of glass fiber mat and/or felt are drawn into the enamel to act as reinforcement.

The overall coating thickness is normally in the range of 2.5–6 mm, depending on the particular specification.

3.2.2.1.2   Asphalt Enamel

Asphalt enamels are manufactured as an end product from the refining of crude oil. It consists of two types:

    a. A straight residue from distillation, which can be of the hard, high melting point.
    b. The "blown" grades that are prepared by partially oxidizing the asphalt base by blowing in air.

Application methods and reinforcement are identical to those used for coal tar enamels.

3.2.2.1.3   Advantages and Disadvantages

Although the two types of enamel appear similar, they are not compatible with each other, and their performance differs markedly.

- Coal tar enamels generally absorb less water than asphalts.
- Minimum holiday susceptibility.
- Good adhesion to steel. Coal tar enamels generally adhere better to clean steel than asphaltic enamels, although this difference is not marked under normal pipeline conditions.
- These coatings should not be used where the pipe operating temperature is likely to exceed 60°C for normal grade coatings. However, higher temperatures can be withstood with higher-grade coatings.
- Good resistance to cathodic disbondment.
- Although both enamels are soluble in chemical solvents, asphalt is dissolved by petroleum products while coal tar is only softened.
- Both coal tar enamel and asphalt coatings are still widely used for submarine lines under concrete weight coatings.
- Low current requirement.
- Limited manufacturers and applicators.
- Health and air quality concerns.
- Change in allowable reinforcement.

3.2.2.1.4   Field Joints and Coating Repairs

The coating of pipeline-weld joints in the field can be achieved by the manual application of coal tar or asphalt or by the application of a self-adhesive laminate tape system. Due to the inherent risks associated with the use of a high-temperature coating under field conditions, this method is rarely used onshore or offshore. Therefore, cold-applied tapes are normally used for field joints and large repairs. Small coating repairs may be carried out by "flaming over" the coating with a gas torch and scraper.

3.2.2.2   Fusion-Bonded Epoxy Coatings

Fusion-bonded epoxy (FBE) coatings have gained wide acceptance in the pipeline industry and can be applied to small and large-diameter pipes. Standards applicable to this type of coating are ANSI/AWWA C213, BGC/PS/CW6, ASTM A972/A972M – 00(2015), etc.

3.2.2.2.1 Description

FBE coatings consist of thermosetting powders that are applied to a white metal blast cleaned surface by electrostatic spray. The pipe is preheated to 230°C, and the quantity of residual heat determines the maximum coating thickness that can be achieved.

Following application, the powder melts, flows, and cures to produce thickness between 250 and 650 microns, following which the pipe is cooled by water quenching. Thickness nearer the maximum is generally required where concrete weight coating is to be applied by impingement methods.

Due to the nature of the coating, strict control of the fusion process is necessary to ensure a satisfactory coating quality (Figure 3.13).

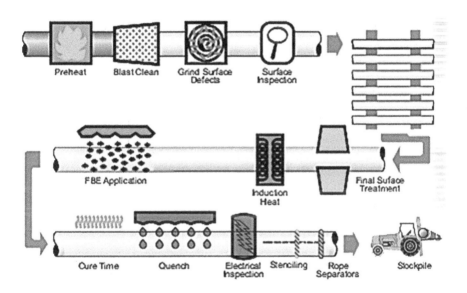

FIGURE 3.13 Fusion-bonded epoxy powder application schematic (Source: *Peabody's control of pipeline corrosion*, 2nd edition, p. 18).

3.2.2.2.2 Advantages and Disadvantages

- FBE coating has been chosen in preference to coal tar enamel to achieve stability at higher temperatures and greater resistance to soil stresses. FBE coatings are suitable for use on pipes with a service temperature of up to 100°C.
- Excellent resistance to cathodic disbondment. Improvements in the performance of FBE coatings have been achieved by carrying out a pre-coating chromate conversion treatment of the pipe immediately after initial surface

preparation. This has led to significant improvements in cathodic disbond-ment and hot water immersion resistance.

- Improvements in adhesion between the concrete and FBE coating can be achieved by increasing the surface roughness.
- Although pipe steel quality has little effect on the application of enamel coatings, special consideration must be given to the pipe steel when apply-ing epoxy powders. Due to the relative thinness of FBE coatings, it is neces-sary after blast cleaning to thoroughly inspect and remove all those surface irregularities, which may cause defects in the finished coating.
- Although the coating is significantly harder than enamels and is able to withstand directs impacts, angular impacts will damage the coating.
- Low current requirement.
- Lower impact and abrasion resistance.
- Excellent adhesion to steel.
- Subject to steel pipe surface imperfections. Due to the relative thinness of FBE coatings, it is necessary after blast cleaning to thoroughly inspect and remove all those surface irregularities which may cause defects in the fin-ished coating.
- Excellent resistance to hydrocarbons.
- High application temperature.
- High moisture absorption.

3.2.2.2.3   *Field Joints and Coating Repairs*
Coating of pipeline-weld joints may be carried out in the field using portable blast cleaning, induction heating, and powder spray equipment.

Alternatively, urethane mastic or laminate tape wrap may also be used.

3.2.2.3   Polyethylene Coatings
Polyethylene coatings are relatively a recent innovation, and consequently less is known about their long-term performance than FBE coatings.

3.2.2.3.1   *Description*
Polyethylene coatings may be applied by one of the following three processes:

   a. Circular or ring-type head extrusion
   b. Powder sintering
   c. Side extrusion and wrapping

In addition, the extruded polyethylene may be applied in conjunction with a primer or adhesive systems.

Polyethylene coating systems favor the use of a high-density polyethylene with either butyl rubber or hot-applied mastic adhesives.

Improved adhesion and resistance to cathodic disbondment can be achieved by priming the pipe surface first with an epoxy-based layer on top of which the adhesive layer and polyethylene coating are applied. This three-layer system is now readily available from various coating plants.

DIN 30670 standard is applicable to polyethylene coatings. The DIN 30670 standard notes that although a thickness of 1 mm is sufficient for corrosion protection, the mechanical load-bearing capacity of the coating may be improved by using a thickness of between 1.8 and 3 mm, depending on pipe diameter.

Sintered polyethylene coatings have recently become available and are produced by pouring polyethylene granules onto preheated steel pipe at 450°C.

Polyethylene coatings should not be used where the pipe service temperature is likely to exceed 65°C. They are resistant to damage during handling and laying. Where concrete is to be used over a polyethylene, consideration should be given to increase surface roughness. Coating disbonding may occur due to the elongation of the pipe caused by a rise in temperature (Figure 3.14).

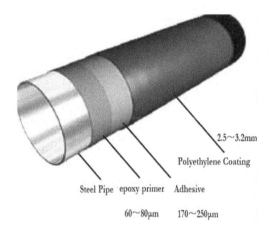

FIGURE 3.14   Schematic diagram of 3LPE (Source: Tong, Shan. 2015. *Cathodic protection.* Training document, Ghana: Sinopec).

3.2.2.3.2   Advantages and Disadvantages

- Polyethylene coatings provide a tough and rugged corrosion protection system for pipelines, with excellent resistance to damage during handling and laying. Although the material can withstand impacts during handling, they do affect the adhesion of the coating, particularly with a hard copolymer adhesive.
- Polyethylene coating should not be used where the pipe service temperature is likely to exceed 65°C.
- Where concrete is to be used over polyethylene, consideration should be given to increase the surface roughness so that a better key between the concrete and the polyethylene coating is achieved.
- Problems have also been experienced with the high-density polyethylene/ rubber adhesive system, where poor quality control during manufacture has led to the variable application of the adhesive. This particular system has shown poorer resistance to the aboveground exposure with embrittlement of the polyethylene and aging of the adhesive. In addition, cases have been

reported of the coating disbonding due to elongation of the pipe caused by a rise in temperature.

- High-tolerance to various kinds of corrosions in different environments.
- Favorable price/performance ratio.
- Favorable insulation performance, immune to long-term dry and immersion conditions.
- High strength, not easily damaged by backfill gravels, of which the diameter is less than 25 mm.
- Anti-disbonding.
- Durable service life, 50 years or more under 60°C.

3.2.2.3.3   Field Joint and Coating Repair

Welded pipe joints are coated using either heat shrink polyethylene sleeves or cold-applied self-adhesive laminate tapes. Field joints are normally prepared by wire brushing, although shot blasting is sometimes preferred for sleeves.

Heat shrink sleeves require careful application to ensure a satisfactory bond is obtained.

Small coating repairs are made with hot melt polyethylene sticks, polyethylene sheet patches, or tape wrap.

3.2.2.4   Tape Wrap

Tape wrap coatings are normally used for field joints, coating repairs, and coating short lengths of pipework in the field. They are not usually applied at the factory due to their inherent susceptibility to damage during transportation and handling. They are normally applied in the field by hand or machine directly from a roll of tape. These tapes are produced in both temperate and tropical grades, and heavy-duty versions are available for use at field joints under hot mastic asphalts or concrete weight coating.

The AWWA C209 standard is applicable to tape wrap coatings.

3.2.2.4.1   Self-Adhesive Bituminous Laminate Tapes

These tapes normally comprise a bituminous adhesive compound layer applied to a PVC backing of varying thickness between 0.08 and 0.75 mm, giving a total tape thickness of approximately 2 mm.

Some tapes are also manufactured with a fabric reinforcement within the bituminous layer to improve impact resistance.

To achieve the required coating thickness, the tape is applied with either a 25 mm or 55% (of tape width) overlap using a purpose-made machine or by hand.

3.2.2.5   Epoxy and Urethane Liquid Coatings

Although the previous sections have summarized the various coating systems generally used for the corrosion protection of pipelines, a number of other paint systems exist which may be used for specific applications where a high degree of chemical or abrasion resistance is required. These paint systems are based on epoxy and urethane resins and may be used to coat short lengths of subsea pipelines and protection structures after fabrication.

3.2.2.5.1   Description

The paint systems normally considered for use on pipelines are the two-pack high-build systems. The preparation of the pipe surface by abrasive blasting is generally required for these coatings in order to achieve a satisfactory bond to the steel. Following this, the paint is normally applied to the required thickness in one coat using airless spray equipment.

Normally, depending on the required service life and conditions, between 1 and 5 mm would be applied.

Both polyurethanes and epoxies can be used with coal tar in order to reduce the material cost.

3.2.2.5.2   Advantages and Disadvantages

- Due to their resistance to abrasion, they are useful for thrust bored pipe installations, and their chemical resistance makes them particularly useful where pipework may be exposed to petroleum or chemical products.
- These coatings are particularly useful for short lengths of tie-in pipework and subsea structures.
- When compared to other coating systems, their main disadvantage is one of cost. In addition, the application is necessary under strictly controlled conditions, and special spraying equipment for hot-applied polyurethanes is necessary. This is because polyurethanes typically contain Iso-cyanates that are linked directly to respiratory problems. In some locations, their use may not be permitted for health reasons.
- Coal tar epoxy and urethane coatings are not recommended for use on pipelines where the service temperature is likely to exceed 80°C.

3.2.2.5.3   Field Joint and Coating Repairs

Coating damage can be repaired by using small quantities of the paint, and field joints can be coated using versions which cure more quickly and which can be hand-painted/troweled on to the pipe surface. As with the pipe coating, field joints require abrasive blasting.

3.2.2.6   Coal Tar Epoxy Coatings

A coal tar epoxy is a black surface protection polymer used on surfaces subjected to extremely corrosive environments. It is a blend of various epoxy resins and coal tar. Coal tar epoxy is made by the conversion of polyamide epoxy with a pitch of refined coal tar. It is mostly used on metal substrates and concrete in offshore, petroleum, and industrial environments. It is commonly used to make high solid coatings or paints to provide moisture protection for underground systems like pipelines, water treatment facilities, clarifiers, and tanks; it is further used in the sewage industry and for prevention from microorganisms. There are different types of paints: two-component paint, three-component paint, etc. The mixture is used in two-component paints.

The fluctuation of temperatures can make the product crystallize. It is stable at room temperature or at least 5°F above the dew point. The environmental conditions affect the drying time of the product (Corrosionpedia 2017).

3.2.2.6.1 Advantages and Disadvantages
- It forms paints or coatings that are smooth in brush, roller, and spray application.
- It bonds well with oily surfaces, hence preferable in garages.
- It forms a good moisture sealing for paints or coatings.
- It produces paints or coatings with abrasion, thermal shock, impact, and chemical resistance, suitable for sustained immersions in saline or freshwater.
- It provides maximum corrosion protection.
- It provides protection against soil stress.
- It provides a semi-gloss with a matte surface finish.
- It provides self-priming and good adhesion in paints or coatings.
- Coal tar epoxy and urethane coatings are not recommended for use on pipelines where the service temperature is likely to exceed 80°C.

3.2.2.7 Mill-Applied Tape Coating Systems

The mill-applied tape systems consist of a primer, a corrosion-preventative inner layer of tape, and one or two outer layers for mechanical protection. Concern regarding shielding of CP on a disbonded coating has led to the development of fused multi-layer tape systems and of a backing that will not shield CP (Sloan 2001).

3.2.2.7.1 Advantages and Disadvantages
- Good adhesion to steel
- Minimum holiday susceptibility
- Ease of application
- Low energy required for application
- Handling restrictions – shipping and installation
- UV and thermal blistering – storage potential
- Shielding CP from soil
- Stress disbondment

3.2.2.8 Extruded Polyolefin Systems

There are two extrusion types available:

a. Crosshead extrusion
b. Side extrusion

Crosshead extrusion consists of an adhesive and an extruded polyolefin sheath. This system is limited to pipe diameters of ½–36 in. (13–900 mm).

Side extrusion consists of an extruded adhesive and an external polyolefin sheath. This system is limited to pipe diameters of 2–146 in. (50–3650 mm).

Extruded polyolefin coatings have performed at higher temperatures. Recent improvements in the adhesive yield better adhesion, and the selection of polyethylene has increased stress crack resistance (Sloan 2001, AWWA C215 1999).

Available with polypropylene for use at higher temperatures (up to 190°F [88°C]), these systems have been used in Europe since the mid-1960s, along with the side extrusion method for larger diameters through 60 in (152.4 cm).

3.2.2.9 Crosshead-Extruded Polyolefin with Asphalt/Butyl Adhesive

3.2.2.9.1 Advantages and Disadvantages
- Minimum holiday susceptibility
- Low current requirements
- Ease of application
- Nonpolluting
- Low energy required for application
- Minimum adhesion to steel
- Limited storage (except with carbon black)
- Tendency for a tear to propagate along the pipe length

3.2.2.10 Dual-Side-Extruded Polyolefin with Butyl Adhesive

3.2.2.10.1 Advantages and Disadvantages
- Minimum holiday susceptibility
- Low current requirements
- Excellent resistance to cathodic disbondment
- Good adhesion to steel
- Ease of application
- Nonpolluting
- Low energy required for application
- Difficult to remove coating
- Limited applicators

3.2.2.11 Multi-Layer Epoxy/Extruded Polyolefin Systems

First introduced in Europe in the mid-1960s as a hard adhesive under polyethylene, followed by the addition of an epoxy primer (FBE or liquid), multi-layer epoxy/polyolefin systems are the most-used pipe-coating systems in Europe. These systems are now available throughout the world (Sloan 2001) (Figure 3.15).

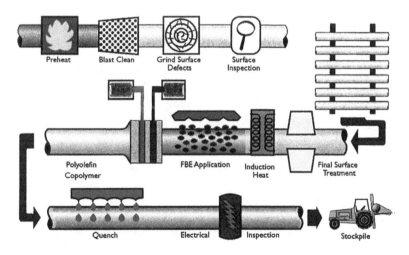

FIGURE 3.15 Three-layer copolymer coating application schematic (Source: *Peabody's control of pipeline corrosion*, 2nd edition, p. 28).

3.2.2.11.1   Advantages and Disadvantages

- Lowest current requirements
- Limited applicators
- Highest resistance to cathodic disbondment
- Exacting application parameters
- Excellent adhesion to steel
- Higher initial cost
- Excellent resistance to hydrocarbons
- Possible shielding of CP current
- High impact and abrasion resistance

3.2.2.12   Elastomer Coatings

A number of elastomer materials are used for coatings. They include:

a. Natural rubber
b. Polychloroprene (neoprene)
c. EPDM (ethylene propylene diene monomer)

Elastomer coatings are generally more expensive than other coating systems described in this book, and so their use is restricted to specific applications where other coatings are inadequate. These include the following:

a. Splash zone coating of risers, where there is a high potential for corrosion, no effective cathodic protection, and a risk of mechanical damage.
b. Pipelines with high fluid temperatures – for example, EPDM is suitable for operating temperatures up to 130°C and polychloroprene coatings for operating temperatures up to 95°C.
c. Applications where corrosion protection and thermal insulation are required, in which the elastomer is normally part of a formulation that includes other materials such as PVC foam, polyurethane foam, or glass spheres.
d. Polychloroprene (PCP) can also be internally applied to spools and can be used as a lining for riser clamps or as an external anti-fouling coating by the application of a 3 mm thick outer layer of PCP containing copper-nickel granules (this system is called cupoprene).
e. Elastomer coating can be used for coating straight lengths of pipe, field joints, and custom parts such as bends and fabricated spools.

3.2.2.12.1   Field Joint and Coating Repairs

Various field joint systems are available for elastomer coatings. They include the use of uncured elastomer that is cured in-situ by electrical heating and the use of formulations that involve PVC foam or mastic materials.

Most elastomer coatings are proprietary systems, and new systems are developed frequently. It is therefore necessary to consult manufacturers about detailed specifications for systems currently available.

3.2.2.13   High-Temperature Coatings

This section deals with coating materials used for onshore and offshore pipelines operating at elevated temperature and gives information on various coating systems available for thermal insulation. Other coatings for equipment and structures at elevated temperatures are also covered.

A key limiting factor for pipelines transporting hot products is the external corrosion coating. A small number of external pipeline coatings are effective at operating temperatures above 66°C (150°F). As the oil and gas industry strives to cut costs and improve profits, high-temperature pipeline coatings become more attractive.

High-temperature coatings are mostly used in the aerospace, manufacturing, military, petrochemical, and power industries for piping, fireproofing, jet engines, offshore rigs, original equipment, and various type of plants/facilities that employ high-temperature processes. One of the largest users of industrial high-temperature coatings is processing facilities such as power plants, petrochemical plants, and refineries.

Coal tar-related coatings could creep to the bottom of the pipe or become brittle and crack. When this occurs the pipeline operator must compensate with more and more cathodic protection. With extruded polyolefin, shrink sleeves, and tapes, high temperatures can cause the adhesive to creep to the bottom of the pipe. This leaves a shell of polyolefin that will shield the cathodic protection. If water penetrates this shell, significant corrosion can occur. At elevated temperatures, FBE can absorb more water than normal, and if under film contaminants are present, blisters can develop, causing the FBE to disbond. Once again, more cathodic protection has to be used to overcome the effects of the disbondment. For these reasons, the operating temperature of the pipeline must be a major factor in the selection of any external pipeline coating (Norsworthy 1996). To prevent the water penetration in the FBE at higher temperatures, several overcoats have been tried. The most successful has been the use of a chemically modified polypropylene (CMPP).

High-temperature coatings for onshore or subsea pipelines that can be used at elevated service temperatures include polypropylene coating, polyurethane elastomer, and foam materials such as syntactic foam. High-temperature coatings for process plants, structures, and equipments include epoxy novolac coatings, silicone coatings, multi-polymeric coatings, epoxy phenolic coatings, and modified silicone coatings

3.2.2.13.1   Standards

Although numerous standards exist which cover the properties of the raw materials for high-temperature and thermal insulation coating systems, no one standard covers their application and repair, with the exception of medium-density polyethylene (MDPE) and polypropylene (PP), which are covered by the DIN 30670 and NFA 49-711 standards, respectively. All coatings detailed are suitable for use at temperatures in excess of 70°C without a concrete weight coating. ASTM D2485, "Standard Test Method for Evaluating Coatings for High Temperature Service", is used to determine the coating's resistance to elevated temperature and a corrosive environment. Resistance to cathodic disbondment is evaluated in accordance with ASTM G42, "Standard Test Method for Cathodic Disbonding of Pipeline Coatings Subjected to Elevated Temperatures".

3.2.2.13.2   Coating Philosophy and Selection

High-temperature and thermal insulation coatings are applied to pipelines to isolate the steel surface from the corrosive environment and provide the pipeline with thermal insulation. Before a choice of the coating can be made, it is necessary to consider the following factors:

- The environment surrounding the pipe
- The pipeline-operating temperature
- Ease of application (yard or site)
- Inspection methods and ease of repair of the defect and pipe joints
- Comparative coating system costs
- Coating properties
- Environment and health constraints
- Degree of thermal insulation required
- Pipeline design life
- Ease of field jointing

Following consideration of the items detailed above, selection of a suitable coating material for the envisaged service conditions may now be carried out. As there are limitations to each type of coating material, an ideal coating system would possess the following characteristics

- Ease of application: the coating should be capable of application in the factory at a reasonable rate, and it should be possible to handle the pipe as soon as possible without damaging the coating.
- The coating must be able to form an excellent bond to the pipe steel with the use of a suitable primer where necessary. In addition, for multi-layer systems, there must be good adhesion between the different coating layers.
- The coating must be capable of withstanding stresses imposed on it by soil/pipe movement.
- The coating must be capable of withstanding impact without cracking or disbonding. Should the outer layer be breached, then the underlying layers should resist water penetration/absorption.
- The coating must be sufficiently flexible to withstand any deformation due to the forces associated with S-lay, reel barge installation, and any expansion and contraction due to temperature change.
- The coating must be resistant to the action of soil bacteria.
- The coating must demonstrate negligible absorption of water and must be highly impermeable to water or water vapor transmission.
- The coating should show no tendency to creep under the prevailing environmental conditions and must have sufficient resistance not to be displaced from the underside of large-diameter pipes.
- The coating must have sufficient inherent roughness or be able to have its surface roughness increased without damaging the coating, to prevent slippage between the coating and a concrete weight coating (when applied).

3.2.2.13.3   Polypropylene Coatings

With the exploitation of increasingly hotter well fluid temperatures, both onshore and offshore, Polypropylene (PP) as a coating material has been increasingly used. In order for the material to maintain its stability at elevated temperatures (>100°C), the polypropylene coating is normally chemically modified.

Although the DIN 30670 standard does not deal specifically with polypropylene coatings, it is frequently referred for their application and testing. The recently published French standard NF A49-711 and ISO 9001:2008 deals specifically with the application and testing of two- and three-layer polypropylene coatings (Chalke and Hooper 1994).

3.2.2.13.3.1   Description   Polypropylene coatings may be applied to line pipe by one of two processes:

- Side extrusion and wrapping
- Circular or ring-type head extrusion

The polypropylene is normally supplied in conjunction with an FBE primer either with or without an intermediate adhesive system depending on whether a two- or three-layer system is being used.

The US pipe coaters favor the two-layer polypropylene system employing a 400–500-micron thickness of FBE and overlaying this with a similar thickness of chemically modified polypropylene.

The European pipe coaters, however, favor the use of a three-layer polypropylene system comprising of a primer layer of FBE some 70–150 m thick, followed by a copolymer adhesive some 230–400 m thick and the chemically modified polypropylene, which can range in thickness from 1 mm to 3 mm (Chalke and Hooper 1994).

Current testing has indicated that a multi-layer coating consisting of FBE 500–750 microns (20–30 mils) thick, topped with 625–1275 microns (25–50 mils) of CMPP, will perform well at internal operating temperatures up to 150°C (300°F) with cathodic protection applied. This system could be applied as a two or three-layer system. The three-layer system may consist of a thick FBE, with 125–250 microns (5–10 mils) of CMPP for a "tie" layer and a thick layer of unmodified polypropylene as a top coat (Norsworthy 1996).

3.2.2.13.3.2   Advantages and Disadvantages   Polypropylene coatings provide a tough and rugged corrosion protection system for pipelines with excellent resistance to damage during handling and laying. This can be an important advantage over "softer" pipe coatings such as FBE coating and coal tar enamel (CTE).

- Although a three-layer polypropylene coating may be able to withstand an impact test of 40 J without obvious external damage, adhesion between the underlying layers may be affected.
- As with polyethylene, where a concrete weight coating is to be applied over the polypropylene, the surface roughness should be increased to provide a

key. This may be easily achieved by sprinkling the surface of the coating with granules of polypropylene immediately after the extruder.

- One of the main advantages of polypropylene is its stability at elevated temperatures, and for this reason, it has been used on projects (both onshore and offshore) having a design temperature up to 120°C and where there is no requirement for thermal insulation.

- As with FBE coatings, pre-treatment of the steel surface using a chromate wash is recommended in order to improve the adhesion of the FBE primer layer for the three-layer polypropylene system. In addition, pipe steel quality is of considerable importance with any raised silvers or other imperfections having to be ground flat after abrasive blasting of the pipe surface.

3.2.2.13.3.3   Field Joints and Coating Repairs   The introduction of polypropylene (PP) as a coating material for line pipe has preceded the development of a satisfactory field joint coating system, and this is particularly true for the three-layer polypropylene system.

Conventional shrink sleeves, while able to bond to the pipe steel and withstand the elevated temperatures, do not form a bond with the polypropylene, and consequently must be considered unsuitable. Although a polypropylene-based shrink sleeve is being developed, indications are that effective bonding between the sleeve and coating at the overlap area is not achieved either.

Cold-applied bituminous tape wrap is considered to be unsuitable also due to the material's inability to resist elevated temperatures.

To date, the most effective joint system is considered to be a two-layer (FBE/PP) type. Although thinner than the pipe coating (approximately 1.8 mm compared to 3 mm), it does have the advantage of being able to resist temperatures up to 120°C. This could be accomplished by applying a two-part epoxy that is partially heat-cured before CMPP powder is applied and melted with a hot, nonstick surface.

A three-layer polypropylene (PP) system has been developed (FBE/adhesive/PP), which will offer a field joint of similar quality to the line pipe coating on some projects which used three-layer PP-coated pipe.

Three-layer polypropylene systems (3LPP) is a multi-layer coating composed of three functional components (FBE/adhesive/PP). This anti-corrosion system consists of a high-performance FBE, followed by a copolymer adhesive and an outer layer of polypropylene (PP), which provides the toughest, most durable pipe coating solution available. 3LPP Systems provide excellent pipeline protection for small- and large-diameter pipelines with high operating temperatures. The FBE component of the 3LPP system provides excellent adhesion to steel, providing superior long-term corrosion resistance and protection of pipelines operating at high temperatures. The superior adhesion properties of the FBE also results in excellent resistance to cathodic disbondment, reducing the total cost of cathodic protection during the operation of the pipeline.

3.2.2.13.4   Polyurethane Elastomer

Polyurethane (PU) elastomers are high-performance solid materials formulated to provide a variety of products with a wide range of physical properties. They are distinctive with respect to their high load-bearing capacity, toughness, and abrasion resistance and typically have a density of 1100 kg/m^3.

3.2.2.13.4.1   Advantages and Disadvantages   PU elastomers have a high natural resistance to fatigue, hydrolysis, and ultraviolet degradation. Combined with excellent resilience, toughness, and abrasion resistance, this makes PU elastomer material an excellent choice for surface coatings in a marine environment.

- Polyurethane elastomers have a maximum service temperature limit of 115°C, and there is no limit on the water depth at which it can be used due to its elastomeric nature. The material can be applied directly to the line pipe as an anti-corrosion coating if required. The material has a thermal conductivity (K) value of 0.22 W/m^{2}K, and as a consequence relatively high thickness is required to obtain low overall heat transfer (U) values, which may make the system excessively expensive or impractical. Due to this, polyurethane elastomers coatings for thermal insulation service normally employ a foam material. Also in terms of material costs, solid PU is slightly less than average when compared to other insulating materials. Polyurethane elastomers coating may be used in conjunction with a concrete weight coating and can be installed by either reel or conventional lay barge.

3.2.2.13.4.2   Field Joint Coating and Repairs   The use of polyurethane elastomer's coating for line pipe does pose some problems in the field joint area. Although castable grades of polyurethane elastomers are available for site use, which are able to withstand a service temperature of 125°C, extreme care has to be taken with the surface preparation of the field/pipe coating interface, if satisfactory adhesion is to be achieved. This also applies to the use of shrink sleeves for field joints.

For thermally insulated pipework, pre-formed half-shells made from a foam material can be used to achieve a low U-value at the field joint. Due to its roughness, repairs to the coating are considered extremely unlikely.

3.2.2.14   Foam Materials

Numerous foam materials are available for use on pipelines requiring thermal insulation depending upon service temperature, water depth, and required U-value, among other things. Foams may be either open or closed cell, although closed-cell materials are preferred for subsea use due to their ability to resist significant water penetration should the outer layer become damaged. The foams most commonly used for pipeline insulation are PU and PVC.

Both of these materials are created synthetically by a process on which a gas (called the blowing agent) is generated within a confined, heated fluid, which expands to form a foam. Blowing agents such as carbon dioxide or Freon-11 have been used to replace the air in the cell structure of the material in order to decrease the thermal conductivity of the material.

Polyethylene foams are flexible, resilient, closed-cell materials. They are chemically inert, and fire retardant grades are available. The foam is available in a range of densities from 25 to 17.5 kg/m^3. The material is manufactured in roll or sheet form and therefore has to be thermo-laminated to produce greater thicknesses (Chalke and Hooper 1994).

Glass foam is a 100% closed-cell, rigid foam material formed by blowing glass with H_2S to produce a foam with a very small cell size. Because it is made from an inorganic material, it is impervious to moisture and non-combustible.

3.2.2.14.1 Advantages and Disadvantages

- PU and PVC foams are each able to withstand a maximum service temperature of 100°C, although both materials are liable to the effects of compressive creep. In addition, it should be noted that the K value of these materials increases with increasing temperature.
- PVC foams vary in density from 200 to 450 kg/m³, and PU foams have a density of 160 kg/m³. PU foam has a maximum water depth limit of 60 m, whereas PVC foam is able to withstand immersion to a depth of 200 m with no ill effects. However, both materials require covering with a solid elastomer to prevent water ingress.
- When used for thermal insulation of pipelines, both PU and PVC foams may be applied directly to the steel surface of the pipe and covered with a solid, water-impermeable elastomer. However, should the outer layer be breached due to impact damage during installation or trawl boards, it is possible for corrosion to occur as foams do not provide any degree of corrosion protection. In order to overcome this, therefore, an anti-corrosion coating of FBE, EPDM, neoprene, PP, or PU is normally applied to the line pipe first, depending upon the operating temperature of the pipeline.
- PE foam, because it is produced in roll or sheet form and has to be thermo-laminated to obtain greater thickness, has not been used for the thermal insulation of pipelines to date.
- Glass foam, on the other hand, has been used on numerous occasions for the production of half-shells for the field joints of thermally insulated pipelines. The complex manufacturing procedure does not lend itself to a foam-in-place application.

3.2.2.14.2 Field Joints and Repairs

Depending upon the type of anti-corrosion coating applied to the line pipe, it may be necessary to apply a similar or compatible system to the field joint area. For the filling of the field joint area, two methods are essentially available. One involves placing a removable mold around the joint and filling the field joint with a liquid polyurethane, which is then allowed to cure. Should a higher degree of thermal insulation be required, then the use of pre-formed foam half-shells may be considered. Conventional practice is then to fill the half-shell gaps with an injected PU elastomer. Additionally, security for the half-shells may be achieved by covering the finished field joint with a heat shrink sleeve.

As with other foam systems, repairs to the coating are considered unlikely because the foam is normally overcoated with a solid elastomer (PU or EPDM). Repairs to the coating system may be feasible but would depend largely on the severity of the damage. Therefore, either any repairs should be performed as a field joint, or the pipe joint should be re-coated.

3.2.2.14.3 Syntactic Foams

Syntactic foams are rigid materials originally developed for deep-water buoyancy. They are manufactured by welding glass microspheres with rigid resin systems. Glass microspheres are used in preference to either polymer or ceramic beads because they do not creep under the combination of pressure and temperature and offer a high insulation factor.

Syntactic polyurethane and polypropylene foams have maximum operating temperatures of 115°C and 120°C, densities of 1100 kg/m^3 and 900 kg/m^3, and K values of 0.22 w/m^2 K and 0.17 W/m^2 K, respectively (Chalke and Hooper 1994).

3.2.2.14.3.1 Advantages and Disadvantages Due to their nature, there is no restriction on the water depth at which they can be used, and both may be used on pipelines, which are to be installed using conventional or reel barge techniques.

- Compared to other available foam insulation materials, the cost of syntactic PU and PP is average. The application of a separate anti-corrosion coating to the surface of the line pipe may not be necessary with syntactic PU because it is an effective anti-corrosion material by itself. Although syntactic PP should be capable of being similarly applied, it is currently applied over the top of an FBE coating.

3.2.2.14.3.2 Field Joints and Coating Repairs As with the conventional foam insulation systems, field joints and repairs to the coating may be achieved, as described in Section 3.2.2.14.2.

3.2.2.15 Epoxy Phenolic Coatings

Epoxy phenolic coatings are classified as either ambient cure, in which the phenolic and epoxy resins chemically react at room temperature, or heat cure, where the coating is exposed to temperatures of 350–400°F to accelerate the cure or activate a catalyst or curing agent in the coating. Epoxy phenolic coating provides chemical, solvent, and temperature resistance and are commonly used for immersion service, tank linings, and high-temperature oil and brine immersion service, for process plants, evaporators, etc., that contain boiling water. Other appropriate applications of epoxy phenolic coatings are used when severe chemical resistance is necessary, but a high degree of flexibility is not (O'Malley 2018).

3.2.2.15.1 Advantages and Disadvantages

- Excellent adhesion properties.
- Temperature resistance up to 400°F. Excellent resistance to boiling water, acids, salt solutions, hydrogen sulfide and various petroleum products.
- Resistance to solvents, chemicals, and abrasion.
- In a fully cured form, they are odorless, tasteless and non-toxic.

- Disadvantages include decreased weatherability and flexibility, relatively slow air curing time and often the necessity of heat curing at relatively high temperatures.

3.2.2.16   Epoxy Novolac Coatings

Epoxy novolac coatings exhibit improved heat resistance because of the presence of aromaticity in their molecular structure, coupled with more cross-linking compared to other epoxies. Novolac epoxies are typically heat resistant up to 350–360°F.

In general, novolac epoxies are known for having greater resistance to oxidizing and non-oxidizing acids, and aliphatic and aromatic solvents compared to other epoxies. These qualities make novolac epoxies an option for applications such as tank linings in contact with high-temperature acidic crude oil. Epoxy novolac coating is ideal for harsh chemical and solvent-resistant applications. It is used in secondary containment, solvent storage, pump pads, trenches, and other high exposure areas (O'Malley 2018).

3.2.2.17   Silicone Coatings

Silicone coatings contain resins that are either pure or hybrid polymers and consist of organic pendant groups attached to an inorganic backbone of alternating silicon and oxygen atoms. The polymer structure provides thermal stability and oxidation resistance. Silicones are essentially transparent to ultraviolet radiation from sunlight (O'Malley 2018).

High-temperature, 100% silicone coatings have a single component and cure by heat-induced polymerization. These thin-film paints dry by solvent evaporation to achieve sufficient mechanical strength for handling and transport. However, the total cure is achieved only after exposure to temperatures in the 350–400°F range. Curing can be achieved as the equipment is returned to its operating temperature. Pure silicone coatings are used on exhaust stacks, boilers, and other exterior steel surfaces at temperatures ranging from 400°F to 1200°F.

3.2.2.18   Modified Silicone Coatings

These high-temperature coatings have lower resistance to elevated temperatures than 100% silicone coatings. Silicone acrylics are single-package air-dry paints that have color and gloss retention to temperatures in the 350–400°F range. Similarly, silicone alkyds are single-package air-dry paints with similar color and gloss retention properties. However, the dry heat resistance of silicone alkyds is limited to about 225°F. Although most high-temperature silicones require ambient temperatures for application, special formulations are available that can be applied to steel up to 400°F.

3.2.2.19   Multi-Polymeric Matrix Coatings

Multi-polymeric matrix coatings have either single- or multi-component inert, inorganic, and resin combinations. Usually, multi-polymeric coatings contain aluminum and micaceous iron oxide flake or titanium. Results from manufacturer studies have revealed anticorrosive performance with single-coat applications (150–200 microns [6–8 mils]) between ambient temperature and 400°C (752°F) in both atmospheric exposure and under-insulation tests (O'Malley 2018) (Table 3.3).

TABLE 3.3
Coating Summary Table

Coating material	Application		Maximum temperature (°C)	Maximum water depth (m)	Cost (*)
	Onshore	Offshore			
Polypropylene	√	√	120	N/A	2
Polyurethane	√	√	115	N/A	4
EPDM (ethylene propylene diene monomer)	√	√	130	N/A	3
Polychloroprene	√	√	95	N/A	3
PU foam	√	√	100	60	2
PVC foam	√	√	100	200	10
Syntactic	√	√	115	300	5
Syntactic PP	√	√	120	N/A	5

(*)
1 = cheapest
10 = most expensive

Source: P. Chalke, J. Hooper. 1994. *Addendum to corrosion protection guidelines.* JPK corrosion protection guidelines, United Kingdom: JP Kenny.

3.2.2.20   Other Coatings

Other coatings to be considered for onshore and offshore pipelines include the use of splash zone coatings (e.g. Monel, anti-fouling paint, nano coating)

3.2.2.20.1   Concrete Weight Coatings

Concrete weight coating (CWC) is a plant-applied coating used to provide negative buoyancy for offshore pipelines or for river or road-crossing applications. Concrete weight coating is used when the stability of the pipeline on the seabed is an issue. The concrete weight coatings also provide mechanical protection against dropped objects and impact by trawl boards. Concrete weight coatings cannot be used on pipelines laid by the reel method. The concrete weight coating is applied to the coated pipe by any of the following methods:

- Impingement
- Extrusion
- Slip forming

The minimum thickness for a concrete coating applied by impingement is about 40 mm, and the minimum thickness applicable by the extrusion process is 35 mm. The two common densities of concrete that are used are 140 and 190 lbs./cu. ft. Higher density is obtained by adding iron ore to the concrete mix. Recently, higher-density iron ore has been used to obtain concrete density ranging from 275 lbs./cu. ft. to

300 lbs./cu. ft. for the Ormen Lange pipeline in the North Sea. An intercoat may be needed over thin coatings such as FBE to protect the thin epoxy coating from damage by the high-velocity-impinged concrete or the concrete extrusion process (Tian et al. 2005).

3.2.2.21   Galvanic Zinc Application

Zinc and zinc alloys are used as a metal coating, which protects the base material, steel, against corrosion. Zinc, in contact with iron (steel), acts as an anode – it dissolves completely and covers with a layer of oxides, inhibiting further corrosion. For zinc coatings, pores are not so crucial.

Galvanizing, while using the electrochemical principle of cathodic protection, is not actually cathodic protection. Cathodic protection requires the anode to be separate from the metal surface to be protected, with an ionic connection through the electrolyte and an electron connection through a connecting cable, bolt, or similar. This means that any area of the protected structure within the electrolyte can be protected, whereas in the case of galvanizing, only areas very close to the zinc are protected. Hence, a larger area of bare steel would be protected only around the edges.

Galvanizing protects the underlying iron or steel in the following main ways:

- The zinc coating, when intact, prevents corrosive substances from reaching the underlying steel or iron.
- The zinc protects iron by corroding first. For better results, the application of chromates over zinc is also seen as an industrial trend.
- In the event the underlying metal becomes exposed, protection can continue as long as there is zinc close enough to be electrically coupled. After all of the zinc in the immediate area is consumed, localized corrosion of the base metal can occur.

The methods of application available are:

a. Zinc metalizing (plating)
b. Zinc-rich paints
c. Hot-dip galvanizing

3.2.2.21.1   Zinc Metallizing (Plating)

Zinc is fed into a heated gun, where it is melted and sprayed on a structure or part using combustion gases and/or auxiliary compressed air (Langill 2006).

Zinc metallizing is a thermal spray application of a zinc and aluminum coating applied to steel. The gas flame bonding of the zinc to the grit-blasted steel is needed to remove any rust or mill scale and prepare the steel for proper adhesion of the zinc. This process protects the steel from corrosion for decades longer than paint alone. This process has been used around the world for 90 years. The steel of every shape and size may be metallized either before or after installation.

Metallizing and hot-dip galvanizing are two zinc-based coatings that protect the steel substrate by a physical barrier and cathodic protection. However, these two coatings are significantly different. Metallizing relies on a mechanical bond between the zinc and the surface of the steel substrate to form a protective coating. Because of this mechanical bond, surface preparation is critical to performance. Hot-dip galvanizing is a total immersion process where the steel element is dipped into a bath of molten zinc.

3.2.2.21.2   Zinc-Rich Paints

Zinc-rich paints contain various amounts of metallic zinc dust and are applied by brush or spray to properly prepared steel.

Zinc-rich paints are those paints that contain a suitably high amount of zinc dust or zinc powder mixed with organic or inorganic binders. Such zinc-rich paints are applied as a topcoat on steel or other metallic surfaces that operate in harsh environmental conditions and that have a continuous risk of corrosion. The zinc dust prevents the metal from becoming corroded by simply sacrificing itself.

Zinc dust is a zinc material in the form of a powder that is also known as zinc powder.

Zinc-rich paints are primarily used to protect metallic surfaces from corrosion by providing cathodic protection at the expense of the zinc contained in these paints.

Zinc-rich paints and coatings are often used with the addition of a primer that acts as a secondary shield to protect the surface from corroding.

3.2.2.21.3   Hot-Dip Galvanizing

Hot-dip galvanization is a form of galvanization. It is the process of coating iron and steel with zinc, which alloys with the surface of the base metal when immersing the metal in a bath of molten zinc at a temperature of around 840°F (449°C). When exposed to the atmosphere, the pure zinc (Zn) reacts with oxygen (O_2) to form zinc oxide (ZnO), which further reacts with carbon dioxide (CO_2) to form zinc carbonate ($ZnCO_3$), a usually dull gray, fairly strong material that protects the steel underneath from further corrosion in many circumstances. Galvanized steel is widely used in applications where corrosion resistance is needed without the cost of stainless steel and is considered superior in terms of cost and life cycle. It can be identified by the crystallization patterning on the surface.

Features of a hot-dip galvanizing coating are (Langill 2006):

- Zinc-iron intermetallic layers
- Harder than the substrate steel
- Zinc patina
- Barrier protection
- Cathodic protection
- Metallurgical bond to the substrate steel
- Paintable
- Edge and corner protection
- Zinc is a natural and healthy metal

According to Langill (2006), the benefits of a hot-dip galvanizing coating are:

- No touch-up required
- High- and low-temperature performance

- Application independent of weather
- 100% recyclable
- Long-term performance in soils, water, and chemical environments
- Maintenance free for 50–100 years in most atmospheric environments (Figure 3.16)

Steel articles are inspected after galvanizing to verify conformance to appropriate specifications. Surface defects are easily identified through visual inspection. Coating thickness is verified through magnetic thickness gauge readings.

Zinc-rich Paints

Zinc Metallizing

FIGURE 3.16 Galvanic zinc applications (Source: Langill, Thomas J. 2006. *Corrosion protection*. Course notes, Iowa: University of Iowa).

3.2.3 CATHODIC PROTECTION

Pipeline corrosion occurs when current flows off a pipeline. Using impressed current system or sacrificial anodes, a cathodic protection system induces current into the ground by using a rectifier and ground bed to stop the flow of current off the pipeline and protecting areas of the pipe not protected by coating.

The basic concepts and requirements for cathodic protection and corrosion protection are outlined here.

3.2.3.1 General

Carbon steel pipelines in the presence of an electrolyte (water) and oxygen will corrode at a significant rate unless measures are taken to prevent it. The two main ways of preventing corrosion are an impermeable coating bonded to the surface of the steel and by a cathodic protection system. The two systems are normally combined for onshore buried pipelines.

There are a variety of coatings available for onshore pipelines. The choice of a particular coating is influenced by pipeline temperature, type of soil conditions anticipated, ability to withstand pipeline installation, and availability in the area. Field-applied systems include tape wrapping, coal tar epoxy wrapping, and painting. Factory-applied systems, which are inherently more attractive due to the more controlled application and ability to grit blast the pipe, include FBE, three-layer polypropylene or polyethylene, or asphalt enamel. FBE is now the most common form of factory-applied coating used for onshore pipelines.

Cathodic protection works on the basis that during corrosion of steel, electric current and difference in potential exist between the corroded area, bare steel, and the ground. A superimposed voltage and current can suppress this reaction and reduce or eliminate corrosion at points where the coating system has broken down or is insufficient.

Before the construction of the pipeline, the following route survey is required:

- Measurement of the electrical resistivity of the soil environment around the pipeline
- Determination of conditions suitable for anaerobic bacterial corrosion
- Determination of various chemical constituents in the soil environment (chlorides, sulfate, sulfides, bicarbonates)

These route surveys will help in the choice of anode materials for cathodic protection.

3.2.3.2   Main Parameters of Cathodic Protection

3.2.3.2.1   Natural Potential

It refers to the corrosion potential obtained without the influence of external current when the metal is buried into the soil.

Factors affecting the natural potential: metal material, surface condition and earth quality, and water content. The natural potential for the coated pipeline generally ranges from -0.4 V to -0.7 V (CSE), but in the rainy season, the natural potential will be more negative. CP design potential for a new pipeline is -0.55 V, on average.

3.2.3.2.2   Minimum Protective Potential

It is the negative potential with the minimum absolute value required for the complete protection of the metal structure (usually -0.85 V with reference to copper sulfate electrode).

3.2.3.2.3   Maximum Protective Potential

The maximum allowable negative potential is slightly higher than the potential that can cause the cathodic disbonding (usually -1.25 V with reference to CSE).

This is a process of disbondment of protective coatings from the protected structure (cathode) due to the formation of hydrogen ions over the surface of the protected material (cathode). Disbonding can be intensified by an increase in alkali ions and an increase in cathodic polarization. The degree of disbonding is also reliant on the type of coating, with some coating affected more than others. Cathodic protection systems should be operated so that the structure does not become excessively polarized, since this also promotes disbonding due to excessively negative potentials. Cathodic disbonding occurs rapidly in pipelines that contain hot fluids because the process is accelerated by heat flow.

3.2.3.2.4   Minimum Protective Current Density

It refers to the protective current density required to slow the corrosion to the lowest extent or to stop the corrosion process. The common unit is mA/m^2.

3.2.3.2.5   Instant Switch-Off Potential

It refers to the ground potential obtained 0.2–0.5 s after the sudden break of the external power or sacrificial anode. As there is no external current flow through the protective structure, the obtained potential is the actual polarization potential without IR drop (voltage drop in the medium).

3.2.3.3   Sacrificial Anode Cathodic Protection System

The potential difference between two different metals of galvanic coupling is the driving force of galvanic corrosion. The strength of corrosion increases as the distance within the galvanic series increases. In the sacrificial anode protection system, it is certain the galvanic distance increases between the anode and the protected structure. The larger the distance is, the more negative potential will be obtained on the protected structure.

Figure 3.17 shows the formation of the sacrificial anode protection system. A complete circuit is formed by connecting the sacrificial anode and the metal to be

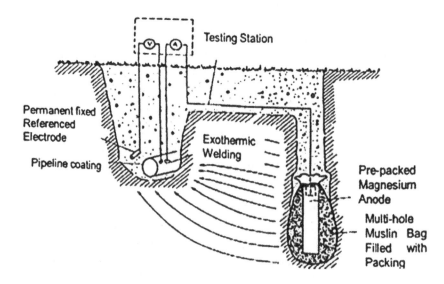

FIGURE 3.17 Composition of sacrificial anode protection system (Source: Tong, Shan. 2015. *Cathodic protection*. Training document, Ghana: Sinopec).

protected. In this circuit, sacrificial anode provides effective current to the protected structure, which can be measured.

In general, sacrificial anodes come in three metals: magnesium, aluminum, and zinc. Magnesium has the most negative electro-potential of the three (see galvanic series) and is more suitable for onshore pipelines where the electrolyte (soil or water) resistivity is higher. If the difference in electro-potential is too great, the protected surface (cathode) may become brittle or cause disbonding of the coating.

Zinc and aluminum are commonly used in saltwater, where the resistivity is generally lower. They are typically used for the hulls of ships and boats, offshore pipelines, and production platforms; in salt-water-cooled marine engines; on small boat propellers and rudders; and for the internal surface of storage tanks.

Requirements for the sacrificial anode system is:

a. Sacrificial anode must have stable and enough negative potential.
b. Anode polarization rate must be small during operation.
c. Theoretical electric capacity of sacrificial anode material must be enough.
d. Self-corrosion rate of sacrificial anode must be very small.
e. Sacrificial anode must have uniform activated dissolution during operation.
f. Corrosive products generated during the operation of sacrificial anode must be non-toxic and harmless with no pollution to the environment.
g. The raw material of sacrificial anode should be abundant and easy to be processed.

3.2.3.3.1   Advantages
- No external power supply required and are easy to install
- No or less interference to nearby structures
- Less managing workload after commissioning
- Economic for minor engineering
- Uniform distribution of protective current

3.2.3.3.2   Disadvantages
- Not suitable for occasions with high resistivity. Ineffectiveness in high-resistivity environments.
- Consume nonferrous metal.
- Commissioning and debugging operation is complex.
- Well coating is required.
- Protective current is nearly non-adjustable.
- Increased weight on the protected structure and increased air and water flow on moving structures such as ships.
- Limited current capacity based on the mass of the anode.

3.2.3.4   Impressed Current Cathodic Protection System

Cathodic protection by an impressed current system requires an input of negative potential between 850 and 950 mV referenced to a copper/copper sulfate electrode.

The negative supply is attached to the pipeline, with the positive supply connected to an anode, buried some distance away from the pipeline. The design of the location and number of these points is dependent on the resistivity of the soil, type, and extent of the pipeline coating and location of potential electrical supplies with which to provide the electricity. Where supplies of external power are unavailable, sacrificial anodes of a more noble metal, e.g. magnesium alloy, are placed on or alongside the pipeline. The anode corrodes preferentially, providing a negative potential to the steel pipeline and helping to protect it from corrosion.

Cathodic protection potentials need to be limited to a negative voltage of 2000 mV to help prevent cathodic disbondment, whereby the gas released by the steel is sufficient to disbond the coating system from the steel surface. Any voids created by these areas are particularly susceptible to corrosion as they are in effect shielded from the anode by the coating and can corrode at a significant rate.

The impressed current cathodic protection system is composed of:

i. DC power supply; auxiliary anode; protected structure; environmental media
ii. Reference electrode
iii. Test station (test post), including current-measuring instruments, potential-measuring instruments, and resistance-measuring instruments
iv. Cable
v. Insulation device
vi. Current loop

Current loop consists of anode bed, cables, DC power supply (current), protected pipeline, and soil. In this loop, the positive pole of the DC power is connected to the

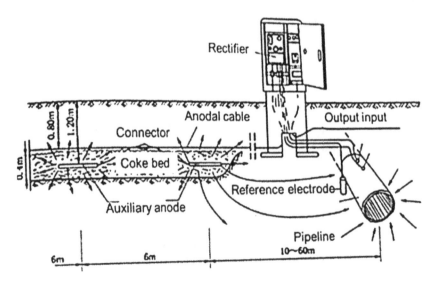

FIGURE 3.18 Impressed current cathodic protection system (Source: Tong, Shan. 2015. *Cathodic protection*. Training document, Ghana: Sinopec).

auxiliary anode, and the cathode pole is connected to the protected structure, sending protective current (cathodic protection current) to protected metal. The effective output current can be tested through the meter in the testing station.

i. Potential loop

The potential loop consists of the pipeline, cables, testing station (voltage meter), reference electrode, and soil. Set a permanent reference electrode in the environment. Protective potential can be monitored and controlled by the potential meter of the testing station. The potential signal of protected structure will be fed back to the testing station or Potentiostat, so that the output DC can be automatically or manually adjusted to the required protection range (Figure 3.18).

3.2.3.4.1 Advantages
- Output current is continuous and adjustable
- Large protection range
- No interference by electrical resistivity of environment
- Economic for a large engineering
- Long service life of the protection devices

3.2.3.4.2 Disadvantages
- External power supply required
- Large interference to nearby structures: possibility of contributing to stray current interference on neighboring structures
- Heavy managing workload

3.2.3.5 Offshore Cathodic Protection

In theory there are several forms of cathodic protection available, but practically most subsea pipelines are cathodically protected using sacrificial anode bracelets of the half-shell or segmented type attached to the pipeline at regular intervals. For subsea structures and platform leg, cathodic protection is by the use of slender stand-off sacrificial anodes or flush-mounted sacrificial anodes which are welded to the structure.

Sacrificial anodes require no source of power.

Principle
- Difference in galvanic series is used
- To provide the electrons to the protected metals
- To make the surface a state of "electron overload"
- To eliminate the potential difference at the surface of the metal
- To stop the electron movement
- To stop the atom from being changed into ions that will flow into the earth (Figure 3.19)

The most common anode configurations are bracelet and stand-off design. However, in some cases, remote anode arrays may be used.

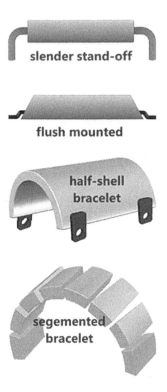

FIGURE 3.19   Typical sacrificial anodes used for cathodic protection.

3.2.3.5.1   Anode Design and Attachment

Once the anode current output and weight criteria are both satisfied, the anode shape and method of attachment to the pipe or structure must be designed to be compatible with:

- Pipeline coating materials, dimensions, and application sequence/methods
- Onshore prefabrication, where applicable
- Offshore and subsea construction activities
- Replacement and inspection activities

One of the key points of design is the need to ensure that areas that should be isolated from the cathodically protected steelwork are not allowed to drain current. The most common anode configurations are bracelet and stand-off design. However, in some cases, remote anode arrays may be used.

3.2.3.5.1.1   Bracelet Anode Shapes   There are two main types of bracelet anode: the twin half-shell and the segmented bracelet. Segment shapes are more easily cast, particularly in large diameters and thicknesses. However, the segmented shape has a more complex steel fabrication, and hence half-shells are preferred when casting is straightforward. Maximum lengths of half-shells are around 1 m, whereas segmented anodes can be cast longer. Anodes can also be "doubled-up" or in "clusters" where greater mass is required. Sketches and dimensions of typical anode shapes can be found in the manufacture's literature.

The steel insert material is used for providing an electrical connection point to the pipe, as well as holding the anode material together. The mass of the insert material, a cage of welded bars, strip, or plate, must be taken from the gross anode mass to give the net mass (W_a). The manufacturer will supply the gross mass of the anode (which includes reinforcement).

Anodes for weight-coated pipelines are usually designed to lie flush with the outer surface of the concrete; otherwise, they are exposed to damage during pipe laying. Recessed anodes can be used where the ratio of current output to weight needs increasing bracelet anodes, for non-weight-coated pipes are normally targeted at each end to ease their passage through handling equipment, stringer, and trenching machine without damage. It is normally small-diameter pipes, i.e. 3-inch outer diameter and below, and consultation with anode manufacturers should be made before designing a CP system.

When pipe laying requires anode installation to be undertaken "quickly" on the lay vessel before passage through the stringer (e.g. reeling of pipe), hinged bracelets are particularly useful as compared with welded half-shells. The utilization factor can be preserved by coating the inside and edge surfaces of the anode.

Anodes for a piggy-backed line may need to allow for the anodes to be recessed such that the piggyback line can lie, supported, inside the anode. Since large pieces of anode material are effectively removed, the mass must be compensated by increasing the length and/or thickness.

Bracelet Anode AttachmentAttachment of the bracelet must ensure two things:

- A permanent, low-resistance electrical contact between anode and pipe
- Prevention of anode movement, which may break electrical connections during installation

3.2.3.5.1.2 Stand-off Anodes (Structures) The design of these anodes is normally more straightforward than for bracelet anodes. For different shapes and sizes, the designer should refer to the manufacturer's data. Anodes for subsea structures are normally placed symmetrically on the structure to give equal load and current distribution. Anodes attached to the top chord should hang vertically down, and those on the top-bottom chord should be attached to outside members where the damage is more likely to be incurred during installation.

Anodes should be placed using sensible judgment regarding areas of high current demand.

3.2.3.5.1.3 Electrical Continuity Electrical continuity between the steel and bracelet anodes (on pipelines) or stand-off anodes (for subsea structures) is via the mechanical attachment where the insert material is welded to the pipeline structure or double plate. In other cases, e.g.:

- Piggy-backed lines which are to be protected by the anodes attached to the larger-diameter flow line

- Steelwork located near the subsea structure which is protected by the stand-off structure anodes

A short length of electrical cables (normally at least two per anode, for redundancy in case of damage) is run between the anode insert material and the pipe wall. One of the most established methods of attaching cables is cad welding, but more recently pin brazing has become an established technique for cable attachment.

3.2.3.5.2 *Anode Materials*

There are several anode materials/alloys in use. Some have a specific use and are not suitable for use in all situations of cathodic protection. The three main metals used as sacrificial anodes are magnesium alloy, zinc alloy, and aluminum alloy.

Magnesium alloys are consumed quickly and only have an 8–10 years life. Magnesium is often used in soil to protect small electrically isolated structures, such as underground storage tanks and well-coated pipelines.

Aluminum alloys are used for subsea pipelines and have three times the current capacity of zinc, the most widely used being Al–Zn–In alloy (Al–Zn–In sacrificial anodes at different concentrations of zinc and indium). Aluminum can be used for a variety of marine applications.

Zinc cannot be used for buried or unburied pipelines above 50°C but is still used for high-resistivity applications. Zinc is often used in marine and soil environments. They are commonly found on boats.

3.2.3.5.3 *Monitoring of Offshore Cathodic Protection System*

For the monitoring of offshore pipeline or offshore structure cathodic protection systems away from landfalls and offshore platforms, there are two methods that can be used. These methods are described below.

3.2.3.5.3.1 Remote Permanent Potential Monitoring Devices   Remote permanent monitoring method is normally used offshore for structures in depths below the diving range and where ROVs can gain no earthing cable contact. This technique transmits the measured structure to seawater potential via an acoustic transponder (ISO15589-1, 2003).

3.2.3.5.3.2 Measurement of Structure to Seawater Potential   The reference electrodes used for monitoring marine cathodic protection systems are the silver/silver chloride/seawater electrode and the zinc/seawater electrode (EN12473, 2006).

Measurement of pipe to seawater potential, current density, or anode current output can be achieved through any of the methods stated below.

3.2.3.5.3.3 Monitoring by Divers   Diving method is expensive and therefore used only for inspection and maintenance works around platforms, wellhead templates, valve assemblage, shallow waters, etc. Divers may be used where an ROV cannot inspect and for very detailed inspections. Information transmitted by divers are not always reliable and reproducible; therefore, such measurement may be carried out with an additional aid of television camera so that the technician on board can record the position and measurement on videotape (Ashworth, Booker 1986).

There are two options for monitoring the potential with the aid of divers:

i. Periodic subsea intervention by divers using hand-held contact probe reference cells to measure structure to seawater potential.
ii. The diver holds the reference electrode as close as possible to the structure and gives the position by radio telephone to those on board where the measurement is recorded. This method is only suitable for a coated structure.

3.2.3.5.3.4 Monitoring by Remote Operated Vehicles   ROVs are frequently used for pipelines and risers. The risers are usually inspected with platform inspection. There are three methods by which an ROV can be used to measure the pipeline protection potential and anode output. These methods are as listed below.

i. Proximity type cell on ROV and hardwire (trailing wire) connected to platform
ii. Two reference cells, one attached as above and one "remote" from pipeline
iii. Two reference cells mounted on ROV

Methods (i) and (ii) are hazardous or need frequent calibration, respectively. Method (iii) measures potential as do (i) and (ii) but can calculate current densities applied to the pipe and output current densities of anodes. The ROV monitoring method can be used for both uncovered and buried pipelines (Baeckmann, Schwenk, and Prinz 1997). For the purpose of cathodic protection monitoring, proximity type instruments are more accurate than contact probes.

3.2.3.5.3.5 Monitoring by Towed Instruments   Monitoring by towed instruments is similar to the ROV technique, where an unpowered "fish" is towed. Although less time consuming, this is also a less accurate technique. It is suitable for buried pipelines.

Close-interval potential surveys provide a nearly continuous plot of the pipeline potential. Towed "fish" or ROVs can be used to carry the monitoring equipment (NRC 1994). ROVs, which follow the pipeline more closely and consistently, are generally most effective. They also can carry video cameras, which reveal even minor coating defects (on pipelines that are not covered with sediments).

3.2.3.5.4 Criteria for Cathodic Protection

For offshore pipelines or structures, steel is under-protected at potentials below −0.80 V (Ag/AgCl/seawater) and below −0.83 V, under protection may occur.

3.2.3.6 Onshore Cathodic Protection

Two types of cathodic protection are used for onshore pipelines (CSA Z662 2015):

a. Sacrificial anode (also known as galvanic anode)
b. Impressed current

3.2.3.6.1 Anode Materials

This section details the various anode materials available in the markets.

3.2.3.6.1.1 Sacrificial Anode Materials For onshore cathodic protection applications, two types of sacrificial anodes materials are used with varying composition to suit a particular application in accordance with ASTM B843. These are as shown in Table 3.4.

Using zinc anodes is restricted to areas where the soil resistivity is less than 5000 ohm.cm, and magnesium anodes uses are normally restricted to areas with soil resistivity less than 10,000 ohm.cm.

High-potential magnesium anodes may be used in soils with higher resistivity due to the greater driving voltage, although the current output will still be restricted.

Sacrificial anodes are normally supplied in a cotton bag filled with a low-resistivity backfill (MESA 2000). This improves the current output of the sacrificial anodes.

Aluminum alloys are normally used for subsea pipelines. For onshore application, aluminum alloys are used for the cathodic protection of pipelines in soils with very low resistivity less than 1000 ohm.cm (Figures 3.20 and 3.21).

TABLE 3.4
Sacrificial Anode Material Specification

Sacrificial anode material	Specification
Magnesium	High purity (standard)
	Galvo-mag or Maxmag (high potential)
Zinc	High purity
	US Military Specification (US Mil Specification)

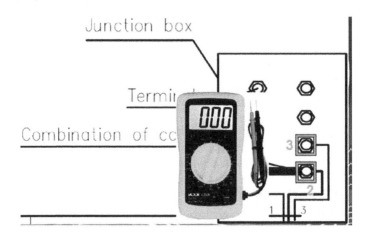

FIGURE 3.20 Galvanic CP (Source: Tong, Shan. 2015. *Cathodic protection.* Training document, Ghana: Sinopec).

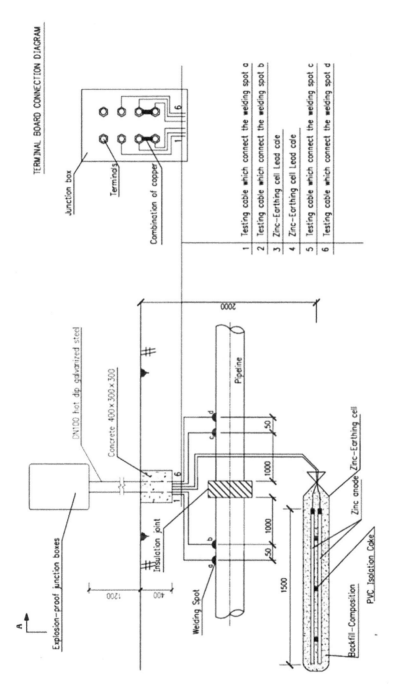

FIGURE 3.21 Galvanic CP and insulation joint (Source: Tong, Shan. 2015. *Cathodic protection.* Training document, Ghana: Sinopec).

3.2.3.6.1.2   Impressed Current Anode Materials   Impressed current anode materials that may be used for onshore applications are:

a. High silicon cast iron anodes (widely used for impressed current applications)
b. High silicon chromium iron anodes (used where there is a possibility of chloride contamination of the ground bed site or anaerobic soil)
c. Graphite
d. Magnetite
e. Platinized titanium (borehole ground beds only)
f. Platinized niobium (borehole ground beds only)

The source of power for an impressed current system could be solar energy battery, wind power generator, transformer–rectifier, thermos batter, etc.

Consideration should be given to isolation joints or insulating flanges at these interfaces:

a. Onshore to offshore pipelines interface
b. Pipeline to structure/terminal interface (Figure 3.22)

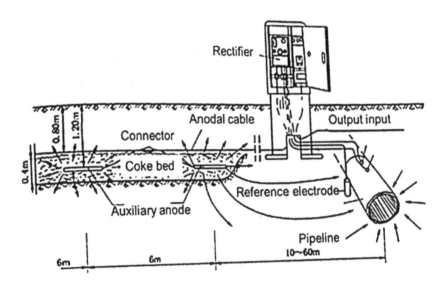

FIGURE 3.22   Impressed current cathodic protection system (Source: Tong, Shan. 2015. *Cathodic protection*. Training document, Ghana: Sinopec).

3.2.3.6.2   Monitoring of Onshore Cathodic Protection System

Cathodic protection monitoring comprises regular inspections performed by pipeline operators to perform measurements at rectifiers and cathodic protection test points along a pipeline (ISO15589-1 2003). The purpose of monitoring is to ensure that the cathodic protection installation remains effective throughout the design life of the structure.

Effective monitoring is by the measurement of structure-to-electrolyte potentials, using a high impedance voltmeter and a reference cell.

This section is mainly about the various methods used for monitoring the onshore cathodic protection system.

3.2.3.6.2.1 Natural Potential Measurement Make sure that cathodic protection is not applied to the pipeline before measurement. For pipeline that has been protected, it is suitable to measure after the power supply is shut down for 24 h. The copper sulfate electrode is placed on the surface-moist soil above the pipe. Make sure the bottom of the copper electrode is well contacted with the soil. Connect the voltmeter, pipeline, and copper sulfate electrode following the method shown in Figure 3.23.

Regulate the voltmeter to the appropriate range. Record pipe-to-soil potential value and polarity marked with the name of potential.

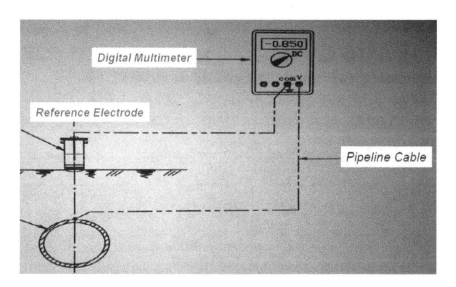

FIGURE 3.23 Natural potential (corrosion potential) (Source: Tong, Shan. 2015. *Cathodic protection.* Training document, Ghana: Sinopec).

3.2.3.6.2.2 Pipe-to-Soil Potential Measurement Monitoring of cathodic protection potentials on onshore pipelines is carried out at the test post facilities installed along the pipeline route using a portable copper/copper sulfate reference cell and a high impedance voltmeter (Figure 3.24).

FIGURE 3.24 Pipe-to-soil potential measurement on Ghana National Gas Company pipeline.

Earth Surface Reference Measurement

It is used for the potential open-circuit or closed-circuit measurement of pipeline natural potential, protection potential, and sacrificial anode.

- Wiring connection is shown in Figure 3.25.
- Use high impedance voltmeter. It is recommended to use a digital multimeter, such as TD-830.
- A reference electrode is placed on the surface-moist soil above the pipe, ensuring well connected with the soil. Use the same method every time to reduce errors.
- The voltmeter should be adjusted to an appropriate range.

Near-Reference Measurement

Near-reference measurement is recommended if IR drop is too larger. As shown in Figure 3.26, the reference electrode is placed close to the pipe wall about 3–5 cm.

Outage Measurement

Outage measurement is used for potential measurement to eliminate the influence of IR drop.

An outage is realized via current interrupter that should be connected in series with the current output terminal of cathodic protection.

During non-measuring period, the cathodic protection station is energized continuously. During measuring of the potential of cathode protection or resistance of the coating, cathodic protection station is energized intermittently, supplying every 12 s with an interval of 3 s. For the same cathodic protection system, interval power should be supplied simultaneously with the synchrony error no more than

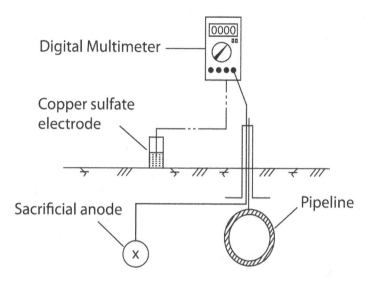

FIGURE 3.25 Surface reference (Source: Tong, Shan. 2015. *Cathodic protection*. Training document, Ghana: Sinopec).

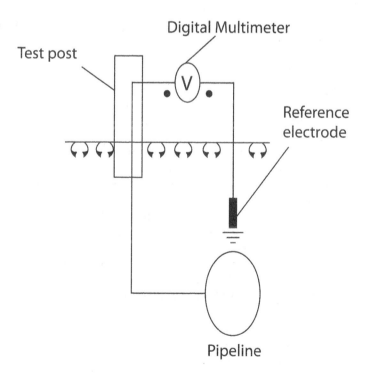

FIGURE 3.26 Near reference (Source: Tong, Shan. 2015. *Cathodic protection*. Training document, Ghana: Sinopec).

0.1 s. The measured potential value obtained via surface measurement with power on is pipeline protection potential when the reference electrode is placed on the pipeline.

Close-Interval Potential Surveys or Close-Interval Surveys

Close-interval potential surveys (CIPS) or close-interval surveys (CIS) are used to determine the level of cathodic protection being experienced along the pipeline by measuring the pipe-to-soil potential and hence corrosion protection levels. There are three types of close-interval surveys: on/off, depolarized, and on surveys.

A measurement is taken by connecting the high-resistance voltmeter negative lead to the reference electrode (half-cell) and connecting the positive lead to the metal being tested. The reference electrode must contact the electrolyte that is in contact with the metal being tested. In soil and freshwater, a copper/copper sulfate reference electrode should be used; in saltwater, a silver/silver chloride reference electrode must be used. To prevent erroneous readings, the voltmeter used must have a minimum of 10 million ohms input resistance under normal conditions; under rocky or very dry conditions, it should have up to 200 million ohms input resistance.

3.2.3.6.2.3 Remote Monitoring Remote monitoring units (RMUs) are available to remotely monitor the output of transformer–rectifier units and drain point potentials. Remote monitoring reduces the cost related to conveying engineers to remote sites. Remote monitoring and wireless transmission of corrosion and cathodic protection data over long distances.

3.2.3.6.3 Criteria for Cathodic Protection

For onshore pipelines/structures, based on a field survey and laboratory test of the soil samples, a decision can be made on the required minimum pipe-to-soil protection criteria. For aerobic soils, the protection criteria would be −0.85 V with respect to $Cu/CuSO_4$ reference electrode. For anaerobic soils containing sulfate-reducing bacteria, the protection criteria would be −0.95 V with respect to $Cu/CuSO_4$ reference electrode (ISO15589-1:2003)

A pipe-to-soil potential of −0.85 V versus a $Cu–CuSO_4$ electrode indicates satisfactory cathodic protection.

The −0.85 criteria, as well as a potential shift of −300 mV versus a $Cu–CuSO_4$ electrode, are used for bare pipelines. It is also suggested that a −100 mV shift of potential in bare pipelines indicates a good degree of cathodic protection.

For high-strength steels with tensile strength in the range of 700–800 MPa, typical protection criteria would be −1.0 V with respect to $Cu/CuSO_4$ reference electrode.

Protective Potential Value (NACE RP 0169-96, SY/T 0036)

- Energized condition: −850 mV
- Polarized potential: −850 mV
- Polarized potential difference: −100 mV

TABLE 3.5
Explanation to the Readings from Cathodic Protection Test

Voltmeter reading	Explanation (what the voltmeter reading indicates)
Greater than −0.88 V	These sections of the pipeline are adequately protected.
−0.85--−0.88 V	The pipe structure at those sections still meets the standard for corrosion protection, but there is not much of a safety cushion. We have to monitor those sections of the pipeline with such reading closely to determine the rate at which the voltage is dropping and plan on adding anodes or performing other work on the system in the not-too-distant future.
Less than −0.85 V	The pipe structure at these sections does not meet the −0.85 V standard for corrosion protection and is out of compliance with regulatory requirements (refer to Clause 5.3.2.1 of ISO 15589-1).
	Possible reason for failure to achieve −0.85 V is
	• Excessively dry soil around the test point.
	• If the pipe's backfill was damp, but −0.85 V reading still could not be measured, then there is the need to research the installation procedures to see if we can discover any clues.
	• Break in a continuity bond or increased resistance between the point of connection and the point of the test due to a poor cable connection.
	• Deterioration of or damage to the pipeline protective coating.
	• Reverse connections at the transformer–rectifier, which is a very serious fault that could result in severe damage to the pipeline in a relatively short period.
−0.4--−0.6 V	It means that the steel pipe has no cathodic protection or that the anodes are completely shot.
	• There is a need for an investigation by a corrosion engineer.

Test Conditions

- IR drop must be eliminated under energized conditions.
- IR drop can be eliminated by applying power-breaking methods.
- Attenuation process during polarization.
- Probe detection method must be applied when there is stray current inter-ference (Table 3.5).

3.2.3.6.4 Steel Casing Pipe Test Station

When a pipeline is encased in concrete or buried in dry or aerated high-resistivity soil, values less negative than the criteria above may be sufficient.

At casing, the test station usually has four wires, two to the casing and two to the carrier pipeline (Figure 3.27).

Under normal conditions, the carrier pipeline should be at a potential more nega-tive than −0.85 V (negative 850 mV) and the casing should be between approxi-mately −0.35 and −0.65 V (a difference of between negative 200 mV and 500 mV).

FIGURE 3.27 Typical casing installation.

3.2.3.7 Satisfying the Current Output Requirement

In order to decrease the amount of sacrificial anode material required for continual protection, submarine pipelines and subsea installations are almost invariably coated with an anti-corrosion system. The coating deteriorates with time, exposing an increasing area of bare metal, so that maximum current output by the cathodic protection system occurs at the very end of the installation design life. The final maximum current output requirement of the cathodic protection system is:

$$I_c = A_c \times i_c \times f_c \qquad (3.4)$$

where

A_c = individual surface areas (m²) of each CP unit
i_c = design current density (A/m²)
f_c = coating breakdown factor (if applicable)
Keys to obtaining enough cathodic protection current are

a. Determine the amount of current required: Theoretical calculations based on coating quality and environment or perform current requirement testing
b. Calculate output expected from the anode and determine the number of anodes required (Figure 3.28)

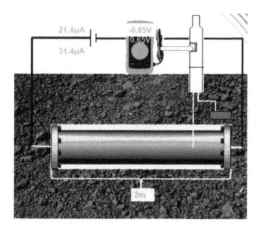

FIGURE 3.28 Current requirement testing (Source: Tong, Shan. 2015. *Cathodic protection.* Training document, Ghana: Sinopec).

3.2.3.8 Additional Requirements

Cathodic protection systems, once installed, require maintenance and inspection in order to identify any areas suffering under protection and can identify areas where increased current demand indicates coating damage. Cathodic protection does not operate when the pipeline is not in contact with the soil, i.e. in casings, in culverts, and on aboveground sections of the pipeline. A good design of the pipeline should eliminate as many of these areas as possible or provide additional corrosion protection.

Use of a cathodic protection system generally requires electrical isolation joints at the ends of the pipeline when connected to aboveground piping, which is normally earthed. For high-pressure pipeline systems, a monolithic joint is normally supplied.

3.2.3.9 Shielding of Cathodic Protection Current

The shielding of cathodic protection current is a common problem that can lead to external corrosion damage and pipeline failures. Coatings with high dielectric strength such as extruded polyethylene, shrink sleeves, and polyethylene tape may lead to the shielding of cathodic protection current if damaged or disbonded.

Improving the quality of the application work can reduce the effects of disbonded and shielding pipe coatings. Alternatively, non-shielding (i.e. fail-safe) coatings such as FBE can be used, especially in high consequence areas such as waterways, populated areas, and environmentally sensitive areas.

Most over-the-line survey techniques will not reliably detect the presence of shielding coatings. In-line inspection and repair is the best way to reduce corrosion failures if disbonded coatings and CP shielding are present.

3.2.4 DESIGN DETAILING

- Install valves that allow for the effective isolation of pipeline segments from the rest of the system.
- Install blinds for effective isolation of inactive pipeline segments.

3.2.5 COMMON CONTRIBUTING FACTORS TO EXTERNAL CORROSION

Excessive heat can cause pipe coatings to soften, flow, or become cracked and brittle, resulting in disbonded and ineffective coating. Soil stresses to backfill weight, soil-induced shear stress applied to the coating due to thermal expansion, pipe settlement, or soil settlement/soil movement can cause disbondment or wrinkling of the coating. Excessive CP current can also cause blisters (in particular, in FBE coating), especially in hot and wet soil environments. The locally increased pH and/or hydrogen molecules being liberated at a holiday in the coating may cause the coating to disbond around a holiday. Cracking or embrittlement of coatings can occur due to prolonged ultraviolet exposure prior to burial. This can happen if the coated pipe is stored outside for long periods. Ultraviolet exposure of fusion bond epoxy coatings may result in chalking and should be evaluated with the manufacturer prior to use. This can also occur at locations where the pipe coating comes aboveground, but it is not protected from the elements. Table 3.6 describes some contributing factors and mitigation of external corrosion of pipelines.

3.2.6 EXTERNAL CORROSION MITIGATION – DESIGN AND CONSTRUCTION

Table 3.7 describes practices for the mitigation of external corrosion during the design and construction of a pipeline's lifecycle.

3.2.6.1 External Corrosion Mitigation – Operation

If failures, direct examinations, and/or risk assessments identify a high susceptibility to inadequate cathodic protection levels or unacceptable external corrosion rates, one or more of the following types of risk mitigation actions shall be initiated:

- External corrosion direct assessment (ECDA) studies of pipelines with suspected external coating degradation to determine CP effectiveness along the length of the pipeline
- In-line inspection of the pipeline to determine the nature and extent of external corrosion damage
- Installation of additional test stations to measure the effectiveness of CP levels at locations where instant "OFF" readings cannot be measured
- Alternative current interruption technology to better assess polarized CP levels
- Upgrading or installation of CP systems to allow for improved current distribution to the pipelines
  - Repair/maintenance of the CP system to ensure that the pipeline is adequately protected and is operated in line with the guidance provided by international practices, such as ISO-15589:2015 or NACE SP0169-2013.
- Repair areas with coating damage or coating disbondment with a suitable system immediately after field investigation (Direct Current Voltage Gradient (DCVG) survey)

Table 3.8 describes the practices for the mitigation of external corrosion during the operating of a pipeline's lifecycle.

TABLE 3.6

Contributing Factors and Mitigation of External Corrosion (CAPP 2018)

Contributor	Cause/source	Effect	Mitigation
Excessive operating temperature	• Coating failure • Coating disbondment	• Water ingress • Cathodic shielding	• Maintain operating temperature below the limit of a coating system • Select the coating system with a temperature greater than the operating temperature
Pipe movement/soil stress	• Excess operating temperature • Operating temperature variation • Improper support	• Coating damage • Water ingress • Cathodic shielding	• Proper pipeline design • Coating selection that meets the design requirements
Ground movement/soil stress	• Unstable soils • Freeze/thaw cycles	• Coating damage • Water ingress • Cathodic shielding	• Route selection • Soil stabilization • Coating selection
Improper handling and backfill	Rock damage	• Coating damage • Water ingress • Cathodic shielding	• Proper construction practices • Coating selection
Poor joint coating	• Poor joint coating selection/incompatible pipe and joint coating • Improper application of joint coating due to inadequate training/supervision/inspection	• Disbonded coating • Water ingress • Cathodic shielding	• Proper design and engineering • Application standards or specification • Trained personnel • Construction quality control • Coatings inspection

(*Continued*)

TABLE 3.6 (CONTINUED)
Contributing Factors and Mitigation of External Corrosion (CAPP 2018)

Contributor	Cause/source	Effect	Mitigation
Improper insulation	• Pipelines without a corrosion barrier between pipe and insulation • Poor joint coating quality that allows water ingress	• Water can enter at holidays and follow the pipe wall • Water can enter the joint area • Outer coating and insulation will shield cathodic protection	• Ensure the coating system includes anti-corrosion barrier • Follow written coating standards or specification to ensure quality work is done on joint coatings • Inject-molded foam at joints rather than half-shells • Employ qualified coating inspectors to ensure the quality of work
Concrete weights and anchor blocks	• Pipelines without adequate coating within the concrete portion • Damaged coating	• Water ingress • Cathodic shielding by the concrete	• Coating must be designed with the consideration for anchor • Coat pipe prior to pouring concrete • Inspect coating prior to installing anchor
Externally weight-coated pipe and rock shielding	• These are not corrosion barriers	• Water ingress • Cathodic shielding	Install holiday-free corrosion barrier applied directly to the pipe
Cased crossings	• Casing in contact with the carrier pipe • Damaged coating	• Cathodic shielding by casing • Insufficient cathodic protection	• Install non-metallic centralizers • Ensure coating is 100% holiday free • Keep water out of the casing

(Continued)

TABLE 3.6 (CONTINUED)
Contributing Factors and Mitigation of External Corrosion (CAPP 2018)

Contributor	Cause/source	Effect	Mitigation
Trenchless crossings – no casing	• Coating damaged during installation	• Water ingress • Cathodic shielding by protective coatings used as rock shields	• Install holiday-free corrosion barrier • Apply cathodic protection • Use abrasion-resistant coating
Soil-to-air interface (risers)	• Damaged coating • Lack of coating	• Coating UV degradation • Coating mechanical damage • Water ingress • Unreliable CP due to intermittent electrolyte	• Proper coating selection • Install coating cap above interface • Inspection and maintenance • Mechanical shielding
Cathodic protection insufficiency	• Cathodic protection system operating below NACE SP0169 criteria	External corrosion at coating defects	Perform the CP system survey and adjust
Cathodic interference	• Foreign cathodic protection systems • AC power lines	Improper cathodic protection	• Properly design cathodic protection system • Proper survey and maintenance
Excessive cathodic protection	• Improperly operated system	Possible coating damage	Perform the CP system survey and adjust

TABLE 3.7

Practices for Mitigation of External Corrosion during the Design and Construction of a Pipeline (CAPP 2018)

Element	Practice	Benefit	Comments
Coating – plant applied	• Select a coating system with design temperature exceeding the operating temperature • Coating selection should consider the type of soil (water, sand, clay, rock)	• Prevents disbondment and cathodic shielding • Minimizes cathodic protection current needed to prevent external corrosion	Monitor operating conditions to prevent exceeding design specifications
Coating – plant-applied Coating –thermally insulated pipe	• Consider a coating system that includes an anti-corrosion barrier between pipe and insulation • Consider water detection wires • Protect and install outer jacket coating system in rocky soils	• Prevents water ingress to the pipe surface • Provides the early detection of jacket breach	• Cathodic shielding may occur due to the insulation • Cannot holiday-check outer coating; therefore, corrosion barrier must be 100% holiday free
Coating – field applied at joints	• Select a joint coating system that considers the current and future operating conditions • Select a joint coating system that is compatible with the pipe body anti-corrosion coating system • Select a joint coating system appropriate for the field construction environment • Use proper surface preparation as recommended by the coating manufacturer • Develop coating application standards or specifications	• Prevents water ingress • Ensures coating system integrity	• Quality control is essential • Applicators must be trained • Applicators must be using the correct equipment and written procedures • Coating inspection should be done to ensure quality and prevent joint corrosion
Joint type	If joints other than butt welds (e.g. zap-lok) are used, consider the effects on cathodic protection	Ensures electrical continuity necessary for the CP system to function, along the full length of the pipeline	Verify by periodic system surveys
Cathodic protection	Install a cathodic protection system	Protects the pipe against corrosion at coating holidays or damage	• Design in accordance with NACE SP0169 • Use proper electrical isolation to avoid current drainage to surface facilities and well casings
Inspection capability	• Install or provide the capability for inspection tool launching and receiving • Use consistent line diameter and wall thickness • Use piggable valves, flanges, and fittings	• Internal inspection using intelligent pigs is the most effective method for confirming the overall pipeline integrity • Proper design allows for pipeline inspection without costly modifications or downtime	Consideration should be given to the design of bends, tees, and risers to allow for the passage of inspection tools

TABLE 3.8

Practices for Mitigating External Corrosion During Operation (CAPP 2018)

Element	Practice	Benefit	Comments
Corrosion assessment	• Understand what type of coatings exist in a pipeline system • Evaluate the operating temperature against coating system design • Assess the potential for cathodic shielding • Re-assess CP system operation subsequent to a line failure or system addition	Understand and document design and operating parameters	Refer to • CSA Z662 Clause 9 • API 579 • ASME B31G • DNV-RP-F101 • NACE SP0502 • ISO 15589 • LI 2189
CP system maintenance	• Perform the annual survey to • Verify sufficient CP current • Check all insulating kits/joints – Check for interference • Check rectifiers periodically and record outputs • Design, install, operate, and maintain the CP system in accordance with NACE SP0169 • Ensure rectifiers are checked routinely to ensure they are operating at the target current output • Reduce unnecessary rectifier downtime due to maintenance activities • React quickly to isolation deficiencies, continuity bonding issues, interference, and other problems to ensure CP systems are functioning properly • Replace depleted ground beds in a timely manner • Upgrade the CP system if more current is needed to provide the proper levels of protection	• Ensure the reliability of the CP system • Enables proof-of-regulatory compliance	• Regulatory requirement • Need to include deactivated, discontinued, or suspended lines • Only abandoned lines should have cathodic protection disconnected

(Continued)

TABLE 3.8 (CONTINUED)
Practices for Mitigating External Corrosion During Operation (CAPP 2018)

Element	Practice	Benefit	Comments
Inspection program	• Develop an inspection strategy • Utilize root cause analysis results to modify corrosion mitigation and inspection programs	• Provides assurance that the corrosion mitigation program is effective • Allows for corrosion mitigation program adjustments in response to inspection results	Refer to CSA Z662 Clause 9 – Corrosion Control
Repair and rehabilitation	• Inspect to determine extent and severity of damage prior to carrying out repair or rehabilitation • Based on inspection results, use CSA Z662 Clause 10.9.2 to determine extent and type of repair required	• Prevents multiple failures on the same pipeline • Prevents recurrence of the problem	• Refer to CSA Z662 Clause 10.10 for repair requirements
Failure analysis	• Recovery of an undisturbed sample of the damaged pipeline • Conduct a thorough failure analysis • Use the results of failure analysis to re-assess the CP system • Measure pipe-to-soil potential at the failure site	Helps understand corrosion mechanisms detected during inspections or as a result of a failure	Adjust corrosion mitigation program based on the results of failure analysis
Leak detection	Integrate a leak detection strategy	Permits the detection of leaks	The technique used depends on access and ground conditions
Management of change (MOC)	• Implement an effective MOC process • Maintain pipeline operation and maintenance records	• Ensures that change does not impact the integrity of the pipeline system • Helps understand and document design and operating parameters	Unmanaged change may result in accelerated corrosion, using inappropriate mitigation strategy for the conditions (outside the operating range)

3.2.6.2 Corrosion Monitoring Techniques

Table 3.9 describes the most common techniques for monitoring corrosion and operating conditions associated with external corrosion of pipelines.

TABLE 3.9

Corrosion Monitoring Techniques (CAPP 2018)

Technique	Description	Comments
Production monitoring	Ongoing monitoring of fluid temperature	Excess temperature may damage the coating
Cathodic protection	Maintain, check, and operate CP system	Refer to NACE SP0169

3.2.6.3 External Corrosion Inspection Techniques

Table 3.10 describes common techniques that should be considered for the detection of external corrosion and coating degradation of pipelines.

TABLE 3.10

Inspection Techniques (CAPP 2018)

Options	Technique	Comments
Cathodic protection effectiveness survey	• Closed interval potential survey • Annual system survey	• Determines adequate protection level • Detects interference • May detect significant coating problem areas • See NACE SP0207 Performing Close-Interval Potential Surveys and DC Surface Potential Gradient Surveys on Buried or Submerged Metallic Pipelines, and NACE SP0286 Standard Practice Electrical Isolation of Cathodically Protected Pipelines • Possible alternatives outlined below
Coating survey integrity	C – Scan Coating Conductance Survey ACVG (pin to pin) and DCVG Coating Survey	• Detailed coating evaluation techniques intended to identify areas of compromised coating • May employ NACE SP0502 Pipeline External Corrosion Direct Assessment Methodology and NACE TM0109 Aboveground Survey Techniques for the Evaluation of Underground Pipeline Coating Condition • Likely does not correlate with areas of extensive external corrosion due to the fact that disbonded coating causing shielding will give erroneous results (may give a false sense of security).

(Continued)

TABLE 3.10

Inspection Techniques (CAPP 2018)

Options	Technique	Comments
Leak detection and monitoring systems (LDMS)	Continuous monitoring of moisture levels	Complete investigations immediately
In-line inspection	Magnetic flux leakage, ultrasonic, and eddy current tools are available. MFL is the most commonly used technique	• Effective method to accurately determine the location and severity of external corrosion • In-line inspection can find external corrosion defects • The tools are available as self-contained or tethered • Can be used for similar-service pipelines to gain a better understanding of issues in an area • The pipeline must be designed or modified to accommodate in-line inspection • May not be effective at risers
Excavation and integrity digs	Physical exposure, inspection, and documenting condition of the coating and the pipe	• Often done on problem areas identified by one of the other options discussed above • If not performed carefully in conjunction with other methods, may give a false indicator of the condition of the coating and pipeline

3.3   INTERNAL CORROSION MITIGATION

3.3.1   CONTRIBUTING FACTORS TO INTERNAL CORROSION IN GAS PIPELINE SYSTEMS

The chemical composition of the fluid that contributes to internal corrosion are:

- H_2O content
- H_2S content
- CO_2 content
- Dissolved solids
- Organic and inorganic acids
- Elemental sulfur and sulfur compounds
- Bacterial and their by-products
- Hydrocarbons
- pH level of the fluid

Physical factors of the fluid stream that contribute to internal corrosion are:

- Temperature
- Pressure
- Velocity
- Vibration
- Entrained solids and liquids
- Deposits
- Flow characteristics and patterns (slug)
- Interaction of all the above and other physical factors

Physical factors of the structure (pipe, vessel, pump, compressor, valves, etc.) contributing to internal corrosion are:

- Material of construction
- Residual or operating stresses
- Design factors
- Crevices
- Deposits
- Surface conditions

Tables 3.11 and 3.12 provide contributing factors to internal corrosion in gas pipelines due to mechanisms and operating practices.

3.3.2 RECOMMENDED PRACTICES FOR MITIGATING INTERNAL CORROSION

Table 3.13 describes the recommended practices for the mitigation of internal corrosion in the design and construction phase of gas pipelines.

Table 3.14 describes the recommended practices for the mitigation of internal corrosion in the operating phase of gas pipelines.

3.3.3 PRACTICES FOR MITIGATING INTERNAL CORROSION – OPERATION

Internal corrosion refers to corrosion occurring on the inside of a pipeline. This type of corrosion often results from the presence of molecules such as carbon dioxide (CO_2), hydrogen sulfide (H_2S), water, organic acids, microorganisms, and other molecules. Typically, these molecules react with the internal pipe surface through anodic and cathodic reactions. The product of these reactions may deposit within the pipe, creating a protective layer that inhibits further corrosion. In other cases, the products do not precipitate and facilitate high rates of corrosion. The rate of internal corrosion depends on the concentration of these corrosive molecules, the temperature, the flow velocity, and the surface material.

For midstream pipelines, a problem called "black powder" occurs because sales gas that is presumed to be dry can still contain some water vapor, which can condense

TABLE 3.11

Contributing Factors – Mechanism (CAPP 2018)

Contributor	Cause/source	Effect	Mitigation
Hydrogen sulfide (H_2S)	• Produced with gas from the reservoir • Can be generated by sulfate-reducing bacteria • Iron sulfide scales tend to dominate when CO_2 to H_2S ratio is less than 20:1 (this limit is supplied as guidance only)	• H_2S dissolves in water to form a weak acidic solution. • General corrosion rates may be slightly decreased with increasing H_2S levels • Hydrogen sulfide can form protective iron sulfide scales • Localized breakdown of iron sulfide scales results in pitting initiation	• Gas sweetening (amine treating) • Cleaning pigging • Injection of corrosion inhibitor • Dehydration • Small amounts of H_2S (i.e. in ppm level) can be beneficial as a protective FeS film can be established
Carbon dioxide (CO_2)	• Produced with gas from the reservoir • Can be introduced as a frac medium	• CO_2 dissolves in water to form carbonic acid • Stability of protective iron sulfide scale may be decreased by an increase in CO_2 • Corrosion rates increase with increasing CO_2 and H_2S partial pressures and temperatures	• Cleaning pigging • Dehydration • Injection of corrosion inhibitor
Bacteria	• Contaminated drilling and completion fluids • Contaminated production equipment • Produced fluids from the reservoir	• Acid-producing and sulfate-reducing bacteria can lead to localized pitting attack • Solid deposits provide an environment for the growth of bacteria	• Cleaning pigging • Injection of biocides and inhibitors • Moisture control
Oxygen	• Ingress from compressors or vapor recovery units (VRUs) • Introduced through endless tubing (ETU) well clean-outs • Ingress during line repairs or inspection • Injection of methanol • Frac fluids saturated with O_2	• Oxygen can accelerate corrosion at concentrations as low as 50 parts per billion • May react with hydrogen sulfide (H_2S) to form elemental sulfur • Typical organic inhibitor effectiveness can be reduced by the presence of oxygen	• Use gas blanketing and oxygen scavengers • Minimize oxygen ingress and/or inhibit the pipeline • Optimize methanol injection and/or use inhibited methanol • Frac fluid to be deaerated or O_2 scavenged

(Continued)

TABLE 3.11 (CONTINUED)
Contributing Factors – Mechanism (CAPP 2018)

Contributor	Cause/source	Effect	Mitigation
Water holdup	• Low gas velocity or poor pigging practices allow water to stagnate in the pipelines • Low gas pressure (e.g. <1000 kpa) may not have the gas density to push water even at higher flow rates • Absence of water separation equipment leads to water-wet pipelines • Deadlegs or inactive service	• Water acts as the electrolyte for the corrosion reaction • Turbulence caused by slug flow regime can accelerate the corrosion rate	• Install pigging facilities and maintain an effective pigging program • Remove the water at the wellsite by separation or dehydration • Control corrosion through effective inhibition • Remove inactive deadlegs • Effectively pig lines as soon as the wells become inactive
Polysulfides	• May be produced with the formation water from sour reservoirs • Polysulfides are water-soluble molecules • Not detected in standard water analysis	• Acidic pH is required for polysulfides to destabilize and precipitate as elemental sulfur • The precipitated elemental sulfur can contribute to accelerated localized corrosion	• Install pigging facilities and maintain an effective pigging program • Implement a corrosion inhibition program
Chlorides	• Produced with the formation water • Can be the result of spent acid returns from well stimulation	• Initiates pitting by disrupting protective scales (more prevalent in sour systems) • Increases the localized pitting rate (increases initiation and acceleration) • Increases the conductivity of water • Increased chloride levels can reduce inhibitor effectiveness	• Dehydration • Injection of corrosion inhibitor

(Continued)

TABLE 3.11 (CONTINUED)
Contributing Factors – Mechanism (CAPP 2018)

Contributor	Cause/source	Effect	Mitigation
Solids deposition	• Includes sands, wax, asphaltenes, and scales • Loose iron sulfide accumulations are commonly formed in sour systems • Can originate from drilling fluids, workover fluids, and scaling waters • May include corrosion products from downhole or upstream equipment • Insufficient gas velocities or poor pigging practices	• Can contribute to under-deposit corrosion • The deposited solids can interfere with corrosion monitoring and inhibition	• Install pigging facilities and maintain an effective pigging program • Initially, use well site separators to tank and truck liquids to minimize the effects of workover and completion activities on the pipeline • Scale inhibition • Install gas conditioning systems
Methanol	• Excessive quantities of injected methanol (methanol injection should be limited to a 1:1 water/methanol ratio or the amount required for hydrate inhibition; methanol can contain up to 70 mg/L dissolved O_2) • Use of uninhibited methanol	• Methanol injection can introduce oxygen into the system • High quantities of methanol may reduce inhibition effectiveness • Methanol can break down protective FeS scales • High quantities of methanol may cause vapor space corrosion	• Avoid over-injection of methanol • Effective pigging and inhibition • Remove free water • The addition of gas dehydration or line heaters can reduce or eliminate the need for methanol usage • Use inhibited methanol
Elemental sulfur	• Produced from a reservoir or formed in the system • Formed due to the reaction of H_2S and oxygen • Without oxygen, thermodynamic instability/solubility due to pressure and temperature change could also generate • Elemental sulfur • Sulfur deposition is more prevalent in liquid hydrocarbon-free systems	• Sulfur deposits can initiate and contribute to accelerated corrosion • Presence of liquid hydrocarbons tends to keep sulfur in solution • Synergistic effects with chloride ion accelerate corrosion	• Install pigging facilities and maintain an effective pigging program • Implement a corrosion inhibition program • Implement sulfur-solvent treatments • Eliminate oxygen ingress

TABLE 3.12

Contributing Factors – Operating Practices (CAPP 2018)

Contributor	Cause/source	Effect	Mitigation
Drilling and completion fluid	• Introduction of spent acids and kill fluids • Introduction of solids • Introduction of bacteria • Introduction of O_2 and CO_2	• Lower pH • Higher chloride concentration, which can accelerate corrosion and reduce corrosion inhibitor dispersibility • Accelerated corrosion due to the breakdown of the protective iron sulfide scales	• Produce well-to-well site separator, tankage, and trucking water until drill and complete fluids and solids are recovered • Supplemental pigging and inhibition of pipelines before and after workover activities
Detrimental operating practices	• Ineffective pigging • Ineffective inhibition • Intermittent operation • Inadequate pipeline suspension practices • Commingling of incompatible produced fluids • Flow back of workover fluids into the pipeline • Deadlegs due to changes in the production or operation of pipelines	• Accelerated corrosion	• Design pipelines to allow for effective shut-in and isolation • Develop and implement proper suspension procedures, including pigging and inhibition • Establish acceptable operating parameters • Test for fluid incompatibilities
Critical gas velocity	Critical gas velocity is reached when there is an insufficient flow to sweep the pipeline of water and solids	• A buildup of water and solids (elemental sulfur, iron sulfide, etc.) causes corrosion • Turbulence caused by a slug flow regime can accelerate the corrosion rate	• Design pipeline to exceed the critical velocity • Establish operating targets based on critical gas velocity to trigger appropriate mitigation requirements (pigging, batch inhibition, etc.)
Management of change	• Change in production characteristics or operating practices • Well re-completions and workovers • Lack of system operating history and practices • Changing personnel and system ownership	Unmanaged change may result in unexpected corrosion	• Implement an effective MOC process • Maintain the integrity of pipeline operation and maintenance history and records • Re-assess corrosivity on a periodic basis

TABLE 3.13

Recommended Practices – Design and Construction (CAPP 2018)

Element	Recommended practice	Benefit	Comments
Materials of construction	• Use normalized EW line pipe that meets standard requirements of standards such as CSA Z245.1 Steel Pipe • Consider using corrosion-resistant non-metallic materials such as HDPE or composite materials • Use Sour Service Steel Pipe for sour gas pipelines, as per the requirements of codes such as CSA Z662	• Normalized EW prevents preferential corrosion of the weld zone • Non-metallic materials are corrosion resistant	• EW pipe should be installed with the seams orientated to the top half of the pipe to minimize preferential seam corrosion • Non-metallic materials may be used as a liner or a free-standing pipeline depending on the service conditions. Steel risers could be susceptible to corrosion
Dehydration	• Install gas dehydration facilities • Ensure dehydration units are operating properly	• Elimination of free water from the system reduces the potential for corrosion	• Consider mitigation requirements for upset conditions
Gas conditioning	• Install inlet separator • Install gas treating (sweetening)	• Inlet separation removes any remaining water and heavy hydrocarbons from the gas stream. • Gas treating is used to reduce the "acid gases" carbon dioxide (CO_2) and hydrogen sulfide (H_2S), along with other sulfur-containing compounds, to sufficiently low levels	
Water removal	• Install water separation and removal	Removal of produced water from the system reduces the potential for corrosion	• Only produced water is being removed therefore pigging and mitigation measures may still be required • Be careful of corrosion due to dew point being reached (condensed water formed may have very low pH due to insufficient buffering capacity)

(Continued)

TABLE 3.13 (CONTINUED)
Recommended Practices – Design and Construction (CAPP 2018)

Element	Recommended practice	Benefit	Comments
Pipeline isolation	• Install valves that allow for effective isolation of pipeline segments from the rest of the system • Install the valves as close as possible to the tie-in point • Install blinds for effective isolation of inactive segments	• Allows for more effective suspension and discontinuation of pipeline segments • Reduces the amount of lost production and flaring during maintenance activities	• Removes potential deadlegs from the gathering system • Be aware of creating deadlegs between isolation valve and mainline at tie-in locations (i.e. install 12 o'clock tee tie-ins, or aboveground riser tie-ins) • Develop shut-in guidelines for the timing of required steps to isolate and lay-up pipelines in each system
Deadlegs	• Design and construct system to avoid or mitigate the effect of deadlegs • Establish an inspection program for existing deadlegs	• Avoid corrosion due to stagnant conditions	• Stagnant conditions lead to accelerated corrosion • For existing deadlegs removal or routine inspection may be required
Pigging capability	• Install or provide provisions for pig launching and receiving capabilities • Use consistent line diameter and wall thickness • Use piggable valves, flanges and fittings	• Pigging is one of the most effective methods of internal corrosion control • Pigging improves the effectiveness of corrosion inhibitor treatments	• Multi-disc/cup pigs have been found to be more effective than ball or sponge type pigs • Receivers and launchers can be permanent or mobile • Use pigs that are properly oversized, undamaged and not excessively worn
Pipeline sizing	• Design pipeline system to maintain flow above critical velocity • For pipelines that operate below the critical velocity ensure corrosion mitigation programs are effective for the conditions	• Using smaller lines where possible increases gas velocity and reduces water holdup and solids deposition	• Design pipeline system to take into account changes in well deliverability • Consider the future costs of corrosion mitigation for oversized pipelines • Consider the impact of crossovers, line loops and flow direction changes
Inspection capability	• Install or provide capability for inspection tool launching and receiving • Use consistent line diameter and wall thickness. • Use piggable valves, flanges and fittings	• Internal inspection using in-line inspection (intelligent pigs) is the most effective method for confirming overall pipeline integrity • Proper design allows for pipeline inspection without costly modifications or downtime	• Consideration should be given to the design of bends, tees and risers to allow for navigation of inspection devices (mandatory for some sour lines per Clause 16 in CSA Z662)

TABLE 3.14

Recommended Practices – Operation (CAPP 2018)

Element	Recommended practice	Benefit	Comments
Completion and workover practice	• Produce wells to surface test facilities until drilling and completion fluids and solids are recovered	• Removal of stimulation and workover fluids reduces the potential for corrosion	• Supplemental pigging and inhibition of pipelines may be required prior to or following workover activities
Corrosion assessment	• Evaluate operating conditions (temperature, pressure, water quality) and prepare a corrosion mitigation program • Communicate corrosion assessment, operating parameters, and the mitigation program to all key stakeholders, including field operations and maintenance personnel • Re-assess corrosivity on a periodic basis and subsequent to a line failure	Effective corrosion management comes from understanding and documenting the design and operating parameters	• Refer to CSA Z662 Clause 9 – Corrosion Control • Define acceptable operating ranges consistent with the mitigation program • Consider the effects of H_2S, CO_2, O_2, chlorides, elemental sulfur, methanol, bacteria, and solids • Consider supplemental requirements for handling the completion and workover fluid backflow
Corrosion mitigation and monitoring	• Develop and communicate the corrosion mitigation and monitoring program to all key stakeholders, including field operations and maintenance personnel • NOTE: Ensure personnel understand their responsibilities and are accountable for the implementation and maintenance of corrosion management programs • Develop pipeline suspension and discontinuation procedures	Allows for an effective corrosion mitigation program	• Refer to Section 3.3 for corrosion mitigation techniques • Refer to Section 3.3.3.7 for corrosion monitoring techniques • Refer to CSA Z662 Clause 9 – Corrosion Control • Number and location of monitoring devices depend on the predicted corrosivity of the system • Process sampling for monitoring of Cl–, pH, Fe, Mn, bacteria, and solids (monitoring of iron–manganese ratio may not be as effective in H_2S system) • Consider provisions for chemical injection, monitoring devices, and sampling points • Establish shut-in guidelines for the timing of requiring steps to isolate and lay-up pipelines in each system

(Continued)

TABLE 3.14 (CONTINUED)
Recommended Practices – Operation (CAPP 2018)

Element	Recommended practice	Benefit	Comments
Failure analysis	• Recover an undisturbed sample of the damaged pipeline • Conduct a thorough failure analysis • Use the results of failure analysis to re-assess corrosion mitigation program	• Improved understanding of corrosion mechanisms detected during inspections or as a result of a failure • Allows for corrosion mitigation program adjustments in response to inspection results	• Adjust corrosion mitigation program based on the results of failure analysis • Some onsite sampling may be required during sample removal (e.g. bacteria testing)
Repair and rehabilitation	• Inspect to determine the extent and severity of damage prior to carrying out repair or rehabilitation • Based on inspection results, use CSA Clause 10 to determine the extent and type of repair required • Implement or make modifications to corrosion control program after repairs and failure investigations, so that other pipelines with similar conditions are inspected and mitigation programs revised as required	• Prevents multiple failures on the same and similar pipelines • Prevents reoccurrence of the problem	• Refer to Section 3.3.4.5 for corrosion inspection techniques • Refer to Section 3.3.4.6 for repair and rehabilitation techniques • Refer to CSA Z662 Clause 10 for repair requirements
Leak detection	Integrate a leak detection strategy	Permits the detection of leaks	The technique used depends on access and ground conditions
Management of change	• Implement an effective MOC process • Maintain pipeline operation and maintenance records	• Ensures that change does not impact the integrity of the pipeline system • Understand and document the design and operating parameters	Unmanaged change may result in accelerated corrosion or using inappropriate mitigation strategy for the conditions (outside the operating range)

and cause corrosion. The corrosion product can then settle as a powder in the pipeline, as well as clog or erode valves and metering equipment. Black powder in gas pipelines is usually corrosion products formed by internal pipeline corrosion but can also be particles from mill scale, weld splatter, formation cuttings, salts, etc., in order to reduce black powder formation in midstream pipelines; the moisture content should be kept low and measures to reduce oxygen contamination should be taken. This should include avoiding using air for drying after hydrotesting and eliminating possible sources of oxygen ingress during operation.

The methods of internal corrosion protection used in the oil and gas industry include:

- Internal coatings
- Gas quality control, e.g. dehydration
- Chemical injection, e.g. corrosion inhibitor and biocide
- Facility maintenance, e.g. cleaning pigging
- Buffering
- Internal cathodic protection (only for the internal protection of tanks)

Internal corrosion of the gas pipe is minimized by reducing the contaminants in the gas entering the inlet of the pipeline. Efforts should be made to enforce strict quality control standards. Product samples should be analyzed on fixed intervals, and solids and liquids taken from the inside of pipelines are checked for corrosive substances such as sulfur compounds and bacteria. The ensuing sections describe the internal corrosion protection methods, as listed above.

One method for reducing the potential for internal corrosion to occur is to control the quality of gas entering the pipeline. Also, by periodically sampling and analyzing the gas, liquids, and solids removed from the pipeline to detect the presence and concentration of any corrosive contaminants, including bacteria, as well as to detect the evidence of corrosion products, a pipeline operator can determine if detrimental corrosion may be occurring, identify the causes of the corrosion, and develop corrosion control measures.

If failures, direct examinations, and/or risk assessments identify a high susceptibility to unacceptable internal corrosion rates, one or more of the following types of risk mitigation actions shall be initiated:

- In-line inspection of the pipeline to determine the nature and extent of internal corrosion
- Review of the existing pigging and chemical program selection and performance
- Assessment of alternative internal corrosion monitoring devices to better assess corrosion activity
- Feasibility study for changing operating conditions to reduce corrosion risk

3.3.3.1   Product Monitoring

Identify key parameters such as pH, temperature, pressure, flow rate, water chemistry, chlorides, dissolved metals, bacteria, suspended solids, chlorine, oxygen, and chemical residuals.

3.3.3.2 Internal Coatings

The principal aim for internally coating pipelines is to reduce pipeline friction and internal corrosion. These characteristics result in lower operating and installation costs, higher product purity, and increased throughput. It has long been established that a protective coating for pipe cores gives countless savings to pipeline operators. The internal coating provides protection against corrosion and abrasion and reduces the cost of scrubbers, strainers, "pigs", and other types of pipeline cleaning services. It ensures product purity, prevents contamination from corrosive products, greatly reduces maintenance and labor costs, and provides protection of the pipe interior against the accumulation of deposits (calcareous or paraffin), and, as has been well established, substantially increased "throughput" of product can be achieved in internally coated pipes.

For both liquid and gas pipelines, the cost of internally coating the pipeline can be justified in most cases only based on reduced operating cost (Kut 1975).

Early field test with gas pipelines showed that, depending on the pipelines and flow characteristics, increase in throughput of 5–10% is possible for 24″ pipeline (Klohn 1959). A potential increase in 1% can justify the cost of internal coating; this measured increase would appear to give economic incentive. However, for most applications, contractual supply or production consideration rules out the case for the application of internal coating based on either increased product throughput or reduced pipeline costs. Therefore, there is a relatively little merit in coating gas pipeline as a means of capacity enhancement.

The pipeline internal surface roughness achievable varies with internal coating material used, but for maximum improvement in hydraulics, the wall roughness should be around 5–10 microns for gas lines (Singh and Samdal 1987).

Known benefits associated with the internal coating of liquid pipelines include reduced maintenance and lower wax depositions (Jorda 1966). Up to 25% reduction in a wax deposition has been shown to be achievable by the application of internal coating (Singh and Samdal 1987).

For liquid systems, the economic benefits are greater for smaller pipe diameters, whereas for gas systems the benefits are greater for larger pipe diameters. In most cases, the benefits are greater for gas than liquid systems.

3.3.3.2.1 Epoxy Pipe Coating

Two-pack epoxy type internal pipe coatings are used in the interior of the pipe used in transmitting dehydrated natural gas, wet gas, crude oil, sour crude oil, salt water, drinking water, fresh water, petroleum products, and numerous chemicals. Such specialized epoxy internal pipe coatings have now been available for a considerable number of years and, because of field experience, can be applied in adequate film thickness, with the required resistance characteristics.

Two main methods for applying internal coatings are:

a. Spraying
b. In-situ coating

3.3.3.2.1.1 In-Situ Coating "In-situ coating" or "in place" coating permits the coating of lines already laid – new or old – and avoids the welding problem. The procedure has been used in the United States for some 20 years and for a considerable number of years in Europe and elsewhere.

Pig Applied In-situ Coating
Internally coating a pipeline using pigs is a more sophisticated recently developed technique in which liquids/paints are moved only with synthetic rubber pigs propelled by compressed air, the material not being enclosed between two balls, cups, or pigs (Kut 1975).

Pig applied in-situ coating is one of the few existing technologies that can deliver a continuous epoxy barrier to long sections of pipe. An example of prevention was executed on a 6-mile long, 18 in. crude oil pipeline, owned by one of the world's largest oil and gas refining conglomerates. The pipeline ran from the coast of the Bohai Sea outside Dongying China to an offshore platform. The pipeline had not experienced any significant corrosion, but the goal was to stop corrosion before it became a problem. After looking at various candidates, the decision was made to internally clean and epoxy-coat the pipeline, which became the solution to the spread of corrosion (Cato 2019).

Surface Preparation and Coating Application
For newly constructed pipelines, abrasive blast cleaning, also known as sandjet method, can be utilized to clean the pipe. More commonly, old or existing pipelines containing corrosion pits and/or scale are chemically cleaned. Both methods are capable of producing SSPC-SP5/NACE No.1 white metal surface.

The cleaning and epoxy coating process consists of five main steps:

1. Hydrocarbon removal
   Removal of all residual oils achieved by batching a carefully chosen detergent between two pigs and shutting it through the pipeline.
2. Acid cleaning and etching
   During this second stage, inhibited hydrochloric acid is batched between two bidirectional pigs and shuttled through the pipeline. Samples are collected from each batch for analysis. The acid is titrated to monitor changes in concentration. As acid reacts with oxidation and scale, the concentration depletes. Once all scales and oxidation are removed, the acid concentration will stop depleting.
3. Water flushing
   Scraper or brush pigs are run with water to remove oxidation and scale that the acid has softened and removed from the pipe wall. Samples are collected from each flush for solid content analysis. The pipeline is ready for additional acid once the water-flush solid content falls below 3%. This ensures that any loose debris will not significantly deplete the acid concentration. The process of acidizing and water flushing will repeat until acid concentrations are no longer depleting, and water samples are clear. Typically, five to six acid batches achieve desired results.

4. Passivating

The passivation stage is a five-step process which stops the development of new oxidation, followed by achieving a desired pH, removal of chlorides, and the use of solvents to expedite drying the pipeline prior to the coating application. The passivation steps are as follows:

- Light hydrochloric acid removes any/all bloom oxidation.
- Phosphoric acid creates a surface for the coating to chemically bond to by converting iron oxide to iron phosphate.
- Rust inhibitor buffers low pH and further passivates the surface.
- Solvent removes moisture and expedites the drying process.
- Dry by purging dehydrated air, with a dew point less than $0°$, through the pipeline.

5. Epoxy coating

The epoxy coating is loaded between two specially designed urethane pigs. Like the way a syringe works, the coating batch works much the same way. As the batch moves through the pipeline, the two pigs are continuously moved closer together. The epoxy coating is extruded onto the pipe wall. The slower the batch travels, the less coating that is extruded. The faster the batch travels, the more coating is applied. This is because the rear pig is pushing into the batch, and the coating is being evenly distributed on the pipe wall. Using a pre-determined drive pressure, the correct velocity is maintained to provide a controlled film thickness. The mil thickness is more accurately achieved and more consistent than other spray-applied methods. Typical specifications may require a minimum of three coating runs. After each coating run, the same dry compressed air continues to lightly blow through the pipeline to ensure total evacuation of the released solvents, which aids in the drying process between coats (Cato 2019).

In-situ Coating Using Rubber Plugs

One basic procedure consists of two specially designed pup joints, fitted with compressed air ports suitably valved, air pressure regulators and bypass units, a coating material inlet port and a quick release, full diameter, and port. Compressed air (or natural gas) is used for driving the specially designed coating plugs.

Two specially designed rubber plugs, comprising rubber disks and rubber bells held together by a mandrel or screwed inserts, contain a batch of coating material between them and are propelled through the pipeline by means of dehumidified compressed air. The fit of the plugs in the pipe is carefully controlled, and the required thin film is deposited behind the rear plug, the thickness being a function of the properties of the coating material, the speed of the coating "train" through the pipe, and the fit of the plugs in the line. The "train" is moved by applying gas or air pressure to the rear plug, back pressure being controlled by maintaining a known constant pressure differential, which controls the speed.

As the paint plugs are propelled by a differential positive pressure, the in-situ coating is therefore forced into close contact with the pipe wall, flowing into surface irregularities.

In-situ Surface Preparation

Surface preparation by abrasive blasting is not employed and is generally carried out by first removing loose scale and rust by passing brushes and scrapers with water and detergent through the line, followed by special techniques of acid cleaning.

Tar, grease, or other contaminants are removed with solvents or if necessary emulsion flushes, and passed as frequently as required, in combination with brushes and scrapers. Variations in the process by applicators include neutralization with ammonium hydroxide after each acid cleaning stage.

In pickling the line with inhibited hydrochloric acid, the concentrations employed and time of surface contact are dependent on the pipe interior condition. Acid residue must be flushed with fresh water, and the surface is then usually lightly phosphate, washed acid free, the water removed, and the pipe finally dried with preferably dehumidified air.

In-situ Lining

The internal pipe coating is then applied in a similar manner – calculated excess volume being introduced, and propelled by dehumidified compressed air. To ensure good wetting and coverage, the coating is reversed. In a multi-coat application, adequate drying time is allowed between coats, depending on the prevailing ambient conditions. Solvent removal is important to ensure freedom from holidays with a dry film thickness varying from 250 microns to 400 microns. The film thickness is closely monitored. The design, fit, construction, speed, and number of passes of the plugs and coating trains to achieve the required film thickness are very much a matter of the expertise and experience of the in-situ coating contractor (Kut 1975).

3.3.3.2.1.2   Treatment of Weld   One of the major problems in internally coating pipes length by length before laying is the subsequent treatment of the weld, where this is essential and couplings cannot be used.

In large-diameter pipes, the welds can be made manually good-though surface preparation is likely to be of lower order. However, this is not possible with smaller diameters.

The remote-controlled auto-detection machine engages the pipe surface at each joint on command and brushes the area of burned coating, caused by welding. On completion of the cycle, the machine automatically cuts off. After a vacuum pipe cleaning tool is run through the pipe to collect dust and slag, a centrifugal impeller replaces the brushes' head, and the tool is re-run through the pipeline for coating operation. During rotation, the extruded brushes clean the weld, as well as both sides. When not rotating, the brushes remain retracted, to avoid damaging the internal pipe coating during travel to and from the work area. To ensure thorough cleaning, the work speed of the carriage is set at about 1 inch per second. A centrifugal impeller applies the coating. To ensure a constant volume of paint during application, the speed of travel of the carriage and the paint flow are synchronized. The total volume depends on the pipe diameter.

3.3.3.2.2   Benefits of Internal Coating to Gas Pipelines

Practical test and experience have shown that the application of internal coating results in the following savings for gas line – service (Kut 1975).

1. Increase in throughput of 4–8%: the increase will be maintained for years, as has been experienced worldwide for many years. However, there is a substantial margin, since it is generally considered that even a 1% improvement in throughput justifies internal coating. For smoothing the internal profile so that gas or fluids flow more readily through the pipe, a thin-film epoxy coating is applied of 1.5–3 mils (37–75 microns) dry film thickness. Application is normally by spraying, following sound surface preparation. This system is essentially used for natural gas pipe insides.

    a. The first major, and now classic, test was in 1958 the "Refugio test". The extensive test was run at Refugio, Texas, on pipelines operated by the Tennessee Gas Transmission Company. This test conclusively proved that the Copon internal pipe coating used increased the capacity of the pipeline – this being the basic proof of increased pipeline capacity, where readings were taken before and after coating.

    b. A 24″ 12-mile pipeline was internally coated. After the lines were coated and dry, the increase in flow efficiency was up to 10% overall. A 4% increase was attributed to cleaning and up to an additional 6% to the coating, depending on the rate of flow.

    c. A further test one year later on the same section of the pipeline showed there has been no apparent deterioration of the flow. Further subsequent tests have indicated that no significant reduction in the capacity confirms the very substantial savings obtained by initial coating. Furthermore, uncoated pipes require frequent cleaning, in contrast to internally coated pipes. These results have been closely paralleled by flow results obtained by many transmission companies since 1958.

    d. Klohn (1959) has fully described the testing procedure for establishing this improved throughput. Reference must also be made to a publication by the American Gas Association (1965). This studies in detail the steady flow in gas pipelines, considering testing, measurement, behavior, and computation. Inter alia, this also refers to internally coated lines. Taylor (1960) has stated that an increase of even only 2% in gas flow can justify the cost of internal coatings. The degree of smoothness of an internally coated pipe is inversely proportional to the friction resistance. The application of such thick films has been repeatedly shown to provide characteristics that are equal or superior to those of new, clean pipes and to regularly pigged pipelines.

2. Easier and faster cleaning of the transmission line after laying, and more rapid drying after hydrostatic testing.

3. Improved flow: in 1846 Jean Louis Poiseuille and a few years later Darcy first developed equations for flow, later refined by Reynolds who conceived "direct" or "streamline flow" and "turbulent flow". Later scientists further

developed the concept leading on to the Reynolds Number and recognition of the adverse effect of pipe roughness on flow in a pipe. The Reynolds Number is an index of the amount of "confusion" inside the pipeline. The greater the Reynolds Number and the rougher the pipe surface, the greater the degree of "confusion"; the smoother the pipe surface, the more orderly the flow pattern and the lower the energy loss due to friction. This theory was confirmed by the Bureau of Mines tests in 1956 and the Refugio test in 1958.

4. Pipe length protected prior to laying: no corrosion which would damage the smoothness and create product contamination.
5. Reduction in paraffin and other deposition – which reduces gas flow.
6. Reduced pumping costs, which are maintained in service.
7. Reduced maintenance: frequency of cleaning is substantially reduced.
   e. A considerable decrease is achieved in the maintenance of coated lines, due to less frequent pigging being required and due to easier cleaning. It was found that in running pigs, about half the pressure was required to move a pig through a coated line as compared to an uncoated line. Similarly, when lines were hydrostatically tested, as compared to an uncoated line. Similarly, when pipelines were hydrostatically tested it was possible to completely dry a 36-inch pipeline with as few as four pig runs. The frequency of pigging varies from pipeline to pipeline, but several major gas transmission companies have provided some data from their experience. This showed that with a coated pipe, pigging was necessary only every 12–18 months. In an uncoated pipe, pigging is normally required about three times a year.
8. Product purity: no contamination from corrosion dust which might block, or damage, applications.
9. Helps pipe inspection: the light reflecting internal coating shows up lamination and other pipe defects.
10. Reduction in friction.

Published data on the loss of pressure in water lines clearly shows that internal coating is not only a vital requirement for protection and maintenance of the installation but that the coating also has a direct effect on losses of pressure and energy.

Condition	Absolute roughness (k_s, mm)
New, bitumen coated	0.01–0.02
New, not bitumen coated	0.04–0.10
Bitumen, partially loosened	0.08–0.10
Light encrustation	0.10–0.20
Cleaned after extended use	0.10–0.20
Overall rusting	0.15–0.40
Chlorinated rubber coating	0.007
Two-component polyurethane	0.001

11. Sound economics

In achieving the above advantages, the initial cost of the coating operation is recovered many times. Even if the diameter of the gas pipeline as installed is adequate for the immediate throughput requirements, internal coating is yet considered advisable in order to allow a margin for the inevitable increased future demand.

An economic higher compromise figure of 3 mls (75 microns) dry film thickness is specified where the gas is mildly corrosive.

3.3.3.3 Chemical Injection

In the general oil and gas industry, the following chemical additives are stored and added to the process to ensure reliable and safe operation.

3.3.3.3.1 Corrosion Inhibitor

These chemicals prevent the corrosion of pipes, pipeline, or tanks. A corrosion inhibitor works by forming a passivation layer on the metal, preventing the access of the corrosive substance to the metal.

A corrosion inhibitor reduces the corrosion rate of a metal exposed to that environment. Inhibition is used internally with carbon steel pipes and vessels as an economic corrosion control alternative to stainless steels and alloys, coatings, or non-metallic composites and can often be implemented without disrupting a process. Corrosion inhibitor or other corrosion protection chemicals such as mono-ethylene glycol (MEG) injected into the system.

The major industries using corrosion inhibitors are oil and gas exploration and production, petroleum refining, chemical manufacturing, heavy manufacturing, water treatment, and the product additive industries (NACE 2016).

3.3.3.3.1.1 Types of Corrosion Inhibitors There are four types of corrosion inhibitors:

a. Anodic inhibitors
b. Cathodic inhibitors
c. Mixed inhibitors
d. Volatile corrosion inhibitors

Anodic Inhibitors
This type of corrosion inhibitor acts by forming a protective oxide film on the surface of the metal. It causes a large anodic shift that forces the metallic surface into the passivation region, which reduces the corrosion potential of the material. This entire procedure is sometimes called passivation (Zavenir 2018). Some examples are chromates, nitrates, molybdates, and tungstates.

Anodic inhibitors are considered dangerous because of their chemical characteristics.

Cathodic Inhibitors
Cathodic inhibitors slow down the cathodic reaction to limit the diffusion of reducing species to the metal surface. Cathodic poison and oxygen scavengers are examples of this type of inhibitor.

Cathodic inhibitors work in two different methods:

a. It may slow down the cathodic reaction itself.
b. It may selectively be precipitating on cathodic regions to restrict the diffusion of eroding elements to the metal surface.

The cathodic reaction rate can be decreased by the use of cathodic poisons. However, it can also enhance the sensitivity of a metal to hydrogen-induced cracking because during aqueous corrosion or cathodic charging the hydrogen can also be absorbed by the metal (Zavenir 2018).

The use of oxygen scavengers that react with dissolved oxygen can also decrease the corrosion rates. Some examples of cathodic inhibitors include:

- Sulfite and bi-sulfite ions that form sulfates when reacting with oxygen
- Catalyzed redox reaction by either cobalt or nickel

Mixed Inhibitors
These are film-forming compounds that reduce both the cathodic and anodic reactions. The most commonly used mixed inhibitors are silicates and phosphates used in domestic water softeners to prevent the formation of rust water (Zavenir 2018).

Mixed inhibitors are film-forming compounds that reduce both the cathodic and anodic reactions. The film-forming solution causes the formation of precipitates on the metal exterior, preventing both anodic and cathodic sides indirectly. Examples of mixed inhibitors include:

- Silicates and phosphates used in residential water softeners to limit the development of rust water.
- In aerated hot water systems, sodium silicate protects steel, copper, and brass.

Volatile Corrosion Inhibitors
Volatile corrosion inhibitors (VCIs), also called vapor phase inhibitors (VPIs), are products moved in a closed atmosphere to the section of corrosion by volatilization from a source. For example, in boilers, volatile compounds such as morpholine or hydrazine are transported with steam to prevent corrosion in condenser tubes by counterbalancing acidic carbon dioxide or by changing exterior pH toward less acidic and corrosive rates (Zavenir 2018).

In closed confined spaces, such as shipping containers, VCI products such as VCI paper, VCI bags, or VCI rust removers are used. When these VCI come in contact with the metal surface, the vapor of these products is hydrolyzed by any moisture to release protective ions.

It is very important, for an efficient VCI, to produce inhibition quickly while lasting for a prolonged period.

Qualities of a VCI product depend on the volatility of its compounds; quick-action-sequence high volatility while providing protection requires low volatility. Examples of volatile corrosion inhibitors include:

- In boilers, volatile mixtures such as morpholine or hydrazine are carried with steam to stop corrosion in condenser pipes.

3.3.3.3.1.2   Applications of Corrosion Inhibitors
- In the oil and gas industry, a combination of film-forming inhibitors and VCIs can be injected into the transmission line to provide 360° protection of the internal pipe walls, protecting against both bottom-of-the-line and top-of-the-line corrosion (headspace).
- VCIs such as Cortec Corrologic VpCI Filler can protect difficult-to-reach areas such as those under disbonded coatings, protect pipeline casing annular spaces, and also help resist bacterial corrosion.
- Hydrotest corrosion inhibitors can be added to the hydrotest water. This hydrotest additive forms a thin film that protects on direct contact but also releases VCIs to help protect the top of the line. After the hydrotest water is drained and the pipeline is capped off, the corrosion inhibitor provides lingering protection.
- Volatile amines are used in boilers to minimize the effects of acid. In some cases, the amines form a protective film on the steel surface and, at the same time, act as an anodic inhibitor. An inhibitor that acts both in a cathodic and anodic manner is termed a *mixed inhibitor*.
- Benzotriazole inhibits the corrosion and staining of copper surfaces.
- Corrosion inhibitors are often added to paints. A pigment with anticorrosive properties is zinc phosphate. Compounds derived from tannic acid or zinc salts of organonitrogens (e.g. Alcophor 827) can be used together with anticorrosive pigments. Other corrosion inhibitors are Anticor 70, Albaex, Ferrophos, and Molywhite MZAP.
- Antiseptics are used to counter microbial corrosion.
- Benzalkonium chloride is commonly used in the oil field industry.
- In oil refineries, hydrogen sulfide can corrode steels so it is removed often- using air and amines by conversion to polysulfides.

3.3.3.3.2   Scale Inhibitor
A scale is a deposit of insoluble inorganic mineral. Common oilfield scales include calcium carbonate and barium sulfate. Scale deposition within processing units such as pipes and heat exchangers obstruct or block fluid flow. A scale inhibitor inhibits scale formation and deposition.

Scale-inhibiting chemicals that are applied up or downhole of the wellhead are, in general, classified into four categories:

- Oil-miscible
- Totally water free
- Emulsified
- Solid

Depending on the mineral content present in the water, duration of the job, and operation needs, the chemical(s) can be applied continuously or in scale squeeze applications.

3.3.3.3.3   Hydrate Inhibitors

See Chapter 7 for information about hydrate inhibition.

3.3.3.3.4   Biocides

These prevent microbiological activity in oil production systems. Uncontrolled bacteria, algae, and fungus activities are problematic in oilfield operations. For example, bacteria activity such that of sulfate-reducing bacteria results in the production of H_2S, which leads to reservoir souring, metal corrosion, health hazards, and clogged filters. Typical uses include diesel tanks, produced water (after hydrocyclones), slop, and ballast tanks.

Biocides are utilized to inhibit and eliminate MIC caused by the corrosive action of microbes. Biocide is injected into the pipeline in the stream of a non-electrolytic carrier. In many cases, the biocide is added to the buffering agent so that only one addition to the gas stream is needed.

Other active agents such as film-formers that aid in forming a passive barrier at the pipe surface and agents that promote the evaporation of electrolytes can also be added. Many of these agents are expensive and, depending on the gas flow at the time of injection or use, may or may not reach the location where the electrolytes and microbes are trapped (Baker 2008).

Biocides discussed here are used in many industries such as the oil and gas industry. Microbial control in the oil and gas industry is primarily practiced to prevent the detrimental effects of microbial growth on production equipment, pipelines, and the reservoir. There are four main groups of biocides:

a. Preservatives (e.g. biocides for liquid-cooling and processing systems, metal-working fluid biocides, and biocides for the oil and gas industry)
b. Disinfectants (e.g. drinking water disinfectants)
c. Pest control
d. Other biocidal products

Antimicrobial agents and corrosion inhibitors are widely used in the oil and gas industry. Treatment chemicals are used in the natural gas industry from well development through transmission and storage of natural gas (Anna, Joanna, and Piotr 2013).

Based on the chemical action of biocides, biocides can be divided into two groups:

a. Substances with oxidizing effect
b. Substances with non-oxidizing effect

The most commonly used oxidizing biocides are chlorine, bromine, ozone, and hydrogen peroxide. Nonetheless, the use of oxidizing biocides is accompanied by these negative effects:

• Interaction with other chemicals (corrosion inhibitors)
• The possibility of interaction with non-metallic substances
• Initiation of corrosion of structural materials

Before each treatment with oxidizing preparations, these effects should be taken into consideration when considering the potential for oxidation, the dose, and the type of treatment (intermittent or constant).

The group of non-oxidizing biocides includes aldehydes (e.g. formaldehyde, glutaraldehyde), acrolein, quaternary ammonium compounds, amines and diamines, and isothiazolones.

Often used in the industry, quaternary ammonium compounds are used as cationic corrosion inhibitors and biocides. The biocidal activity of these substances is to dissolve the lipid cell membrane, which leads to the loss of the cell contents of the microorganism.

Quaternary ammonium cations, also known as QUATS, are positively charged polyatomic ions of the structure NR_4^+, R being an alkyl group or an aryl group. QUATS prevent the formation of polysaccharide secretions during bacterial colonization, thus showing antibacterial activity. QUATS are used in closed systems and gas manifolds. On the other hand, they are not used during the exploitation of oil because they may adversely affect the permeability of the crude oil deposit. Furthermore, they are not compatible with oxidizing agents, especially the chlorates, peroxides, chromates, or permanganates. Most of these compounds are readily biodegradable. Benzalkonium chloride is a common type of QUAT salt used as a biocide, a cationic surfactant, and a phase transfer agent.

Isothiazolones is another type of biocides. They are fast-acting biocides inhibiting the growth, metabolism, and biofilm formation by algae and bacteria. They are used in combination with other biocides, or, individually, typical aqueous solutions of chloride- and methyl-derivatives of these compounds are used. Isothiazolones are used only in an alkaline medium; at pH < 7 they lose biocidal properties. Moreover, these compounds can be used in combination with other chemicals without changes in performance (Anna, Joanna, and Piotr 2013). An exception is the environment containing hydrogen sulfide, which causes the deactivation of isothiazolones. The main application areas of isothiazolones are coolants and cooling and lubrication fluids.

Another compound is *glutaraldehyde* (pentane-1,5-dial). This is the most common component of commercial biocides with powerful antibacterial and antifungal activity. An important advantage of this compound is the possibility of use in a wide range of temperatures and pH, as well as solubility in water. Glutaraldehyde does not react with strong acids and alkalis but reacts violently with ammonia- and amine-containing substances, causing exothermic polymerization reaction of an aldehyde, and thus its deactivation. It is not sensitive to the presence of sulfides and tolerates high-salinity environments (Anna, Joanna, and Piotr 2013).

Another biocide used in the industry is *Tetrakis (hydroxymethyl) phosphonium sulfate (THPS)*. It is water-soluble, ionic-biocide-destroying bacteria, fungi, and algae in industrial cooling installations and process water tanks. It is characterized by low toxicity and interacts with other chemicals used in aqueous environments; a particular advantage of this compound is its ability to remove residual iron sulfide in pipelines. THPS biocide is a kind of environmental-friendly water treatment microbiocide that is made of Tetrakis (hydroxymethyl) phosphonium sulfate (THPS 75%) solution. It can withhold sulfate-reducing bacteria (SRB), most of aerobic bacteria including microorganisms that form biofilm in enhanced oil recovery process, production and other supporting systems such as water injection equipment, well water disposal facilities, water-holding tanks, recirculating water treatment systems, and pipelines. THPS biocide is also effective in controlling microbial growth in drilling muds and stimulation fluids for oil and gas wells. Tetrakis (hydroxymethyl) phosphonium sulfate is characterized by its low solidity point and good stability. THPS 75% solution can easily dissolve in water and can be preserved for a long time (Anna, Joanna, and Piotr 2013).

Structural formula:

In gas storage systems, the traditional treatment is with biocide and/or corrosion inhibitor. The chemical is mixed with water or a suitable solvent, pumped into the storage well, and followed with gas. This method assumes the gas will displace the chemical mixture into the formation to control bacterial growth. Although negative bacteria cultures have been noted with this type of treatment, there is concern that the biocide stayed in the bottom of the well or near the wellbore and was not carried far into the formation where bacterial growth still continued.

Side effects of biocide treatments are foaming, emulsions, and expensive equipment costs for individual well treatments.

3.3.3.3.5 Antifoam

It is a chemical additive mixed with industrial process liquids so that the formation of foam in the industrial process liquids can be avoided. Some of the commonly used defoamers or anti-foaming agents are insoluble oils, glycols, polydimethylsiloxanes, silicones, octyl alcohol, aluminum stearate, and sulfonated hydrocarbons are used as antifoam/defoamers (Corrosionpedia 2017).

Antifoams prevent the formation of foam in oil processing by reducing liquid surface tension. It is used especially in units such as separators where foaming impedes the effective separation of gas from the liquids. Antifoam is used to eliminate foam that may cause overflowing, clogging, corrosion, or electrical shortout. If foam is produced in any industrial process liquid, it can cause serious problems such as defects in the surface coatings due to the formation of air bubbles in the surface coatings. Because of this, unevenness occurs and a smooth surface coating is not achieved.

3.3.3.3.6 Drag Reducers

They improve flow in pipelines. During fluid flow the portion of the fluid in contact with the wall is slower than the portion in the center. This leads to the formation of turbulent eddies, which cause a drag. A drag reducer suppresses the formation of turbulent eddies reducing the drag, and subsequently the frictional pressure drops in the pipeline.

Drag reducers, also known as flow improvers of the long-chain polymer type, have been known since 1946, when B. A. Toms, a British chemist in London, first undertook experiments. He found the drag reduction phenomenon while studying the characteristics of liquid solutions in a turbulent flow.

Energy must be applied to the fluid being transported through a pipeline. The energy moves the fluid but is lost in the form of friction as the fluid moves down the pipeline.

Flow improver technology can reduce the energy lost due to friction or drag by more than 50% in most cases and increase flow rates by up to 100%. The performance depends largely on the properties of the fluid being transported and the condition of the pipeline.

Erosion can occur because of the contact between the pipe wall and the flow of gas or hydrocarbon liquid. The most prominent factors contributing to the levels of erosion include flowrate and the level of contamination present in the gas or fluid. For example, the greater the volume of contamination within the pipeline, the greater the risk of erosion. Additives such as drag reducers/flow improver have the ability to prevent flow-induced localized corrosion (FILC) or erosion corrosion, internal corrosion, and corrosion of the metal wall. Drag-reducing additives reduce freak energy densities to values significantly below fracture energies of protective layers and hence inhibit the initiation of FILC (Schmitt and Bakalli 2008).

3.3.3.3.6.1 Drag Reduction "Drag" is a term that refers to the frictional pressure loss per length of the pipe that develops when a fluid flows in a pipeline. Drag increases with the increasing flow velocity. Drag reduction is the proportional decrease in this frictional pressure drop achieved with the addition of very small amounts of a specialty chemical that acts as a drag-reducing agent, also called a drag-reducing additive or a flow improver (Dean 1984).

Drag reduction, as defined by Savins, is the increase in pumpability of a fluid caused by the addition of small amounts of an additive to the fluid.

The effectiveness of a drag reducer is normally expressed in terms of percent drag reduction.

At a given flow rate, percent drag reduction is defined as:

$$\%D.R = \frac{P - P_p}{P} \times 100 \qquad (3.5)$$

where

P = base pressure drop of the untreated fluid

P_p = pressure drop of the fluid containing the drag-reducing polymer

$D.R$ = drag reduction

Generally, the increased pumpability is used to increase the flow rate without exceeding the safe pressure limits within the flow system. The relationship between the percent drag reduction and the percent throughput increases can be calculated using the equation below:

$$\text{Percent throughput increase} = \left[\left(\frac{1}{1 - \frac{\%D.R}{100}} \right)^{0.55} - 1 \right] \times 100 \qquad (3.6)$$

where

$\% D.R$ = the percent drag reduction as defined in Equation (3.6).

Equation (3.6) assumes that the pressure drop for both the treated and the untreated fluid is proportional to the flow rate raised to the 1.8 power.

In most petroleum pipelines, the flow through the pipeline is turbulent. Turbulent flow is characterized by the irregular, random motion of fluid particles in directions transverse to the direction of the main flow. The flow is unstable. Turbulent eddies are generated at the pipe wall and move into the core of the pipe. More energy is required to transport fluid at a given average flow velocity in the turbulent flow because not all of the energy is dissipated in the formation of eddy currents.

In most cases, a general family of polymeric chemical additives called drag reducers can decrease this turbulent energy loss (Figure 3.29). Generally, the more turbulent the flow, the more effective the drag reducer becomes, and, consequently, more efficient energy utilization can be achieved (Figure 3.30).

FIGURE 3.29   Flow regimes in a pipeline.

Turbulence is first formed in the buffer zone, and drag reducers are most active in the buffer zone. The actual performance depends on the hydraulic characteristics of the pipeline and the physical properties of the liquid.

3.3.3.3.6.2   Pipeline Drag-Reducing Additive   A pipeline drag-reducing additive is typically a high-molecular-weight hydrocarbon polymer suspended in a dihydrocarbon solvent. When mixed with a fluid such as crude oil or refined petroleum products in pipelines, it changes the flow characteristics and reduces the flow turbulence in the pipeline.

The strength of the turbulent eddy currents at the pipe wall is reduced by the addition of a drag reducer. Some believe that PDRA absorbs part of the turbulent energy and returns it to the flowing stream. By lowering the energy loss or drag, the PDRA allows the pipeline throughput to increase for a given working pressure, thereby increasing normal pipe capacity or throughput or to operate at a lower pressure for the same throughput, thereby decreasing operating cost.

Pipeline drag-reducing additives do not work by being absorbed into or coating pipelines, but rather they are dissolved into and become part of the fluid. The PDRA is highly susceptible to shear stresses in a pipeline system and thus loses its

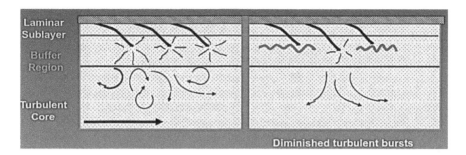

FIGURE 3.30 Drag reduction mechanism (Source: Laura Thomas, Tim Burden. 2010. *Heavy oil drag reducing agent (DRA): increasing pipeline deliveries of heavy crude oil.* UK: ConocoPhillips specialty products Inc.).

effectiveness readily. This happens mainly as the internal shear stresses of the pipeline flow and geometry break the PDRA long-chain molecules into smaller pieces. Thus, the PDRA must be added continuously to the pipeline fluid to maintain the desired level of drag reduction.

3.3.3.3.6.3 Injection of Flow Improvers A compact, skid-mounted chemical injection module consisting of injection pumps and positive displacement flowmeters with totalizers and miscellaneous instrumentation, is used to inject the polymer. The polymer is viscous, so an inert gas, usually nitrogen, is used to help transfer material from tank to pumps (Figure 3.31).

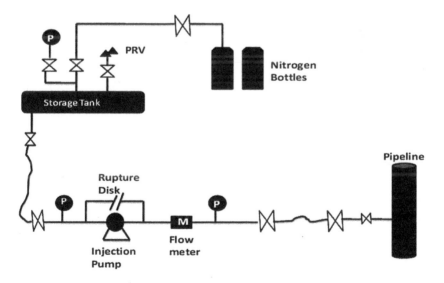

FIGURE 3.31 Injection system for drag reducer/flow improver (Source: Dean, Hale. June 1984. "Special Report-Slick Way to Increase Capacity". *Pipeline & Gas Journal* 17–19).

Economics of Flow Improvers

The economics of using a drag reduction additive depends on:

- How much drag reduction or flow increase is needed?
- The characteristics of the crude oil/petroleum products being transported
- The pipeline configuration

Limitations

For the drag-reducing additives to work:

- The flow through the pipeline must be turbulent.
- The additives are usually effective up to more than 70% flow increase.
- For flow increase above 50% a combination of new pumps or loop sections, together with additives may be required.

Applications of Flow Improvers

Initially it was thought that only very large-diameter lines delivering huge volumes could benefit from the use of drag reducers. The test has demonstrated that smaller crude and product lines can also benefit. The use of flow improvers in pipelines results in the following advantages:

a. Flow increase
- Increase the flow rate of one or many pipeline segments
- Increase the tanker loading rate
- Increase floating storage offloading (FSO) loading rate

The most common use of drag-reducing additives is to increase pipeline flow rates in a system that is at capacity. In order to deliver more hydrocarbon fluids in the system, DRA is injected at each segment to increase the overall pipeline capacity (Figure 3.32).

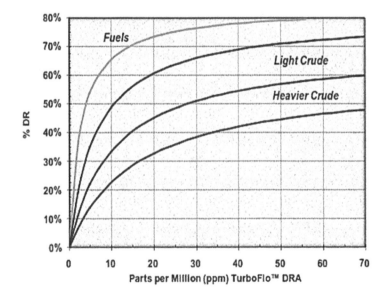

FIGURE 3.32   Typical Turboflo DRA performance.

b. Reduces capital expenditure

DRA usage is a very cost-effective versus capital-intensive mechanical expansion (limits the need for additional pumping facilities or looped pipe segments). In addition to no capital requirements to achieve the extra flow, the desired incremental increase is nearly instantaneous upon injection. This type of application is utilized in many oil-producing regions. The cost is minimal per incremental barrel produced.

c. Pressure reduction
  • Reduce operating pressure to handle corrosion problems
  • Reduce operation pressure due to maintenance
  • De-bottleneck connected platform infrastructure

Pipeline operations depend on pumping pressure as their lifeline to deliver flow capacity. Due to aging systems, corrosion, abrasions, or a pipeline bottleneck, pressure constraints become a pipelines Achilles heel. Allow continuous operation within the constraints of maximum allowable operating pressures without sacrificing capacity.

Drag-reducing additives are used to manage pressure constraints and/or enable typical capacity at lower pressures.

d. Energy savings

Drag-reducing additives are commonly used to maintain a pipeline's flow rate while bypassing a pumping station that is down temporarily for service or repair. However, drag-reducing additives can be used as a permanent pump replacement.

This inventive approach has allowed the operator to bypass half of the pumping stations on the system and significantly lower overall energy use. This "energy hedge" delivers significant seven-figure dollar savings each year, reduces maintenance and repairs costs, and provides a "green" alternative to energy usage (Table 3.15).

3.3.3.3.6.4 Effects of Drag Reducer on End-Use Equipment Of major importance is the effect of drag reducers on the equipment and operations that ultimately the end users must be responsible for. Previous works on drag reducers have been accomplished by Conoco on its own Conoco Drag reducer (CDR). Both Conoco and Trans-Alaska Pipeline System (TAPS) ran a set of bench tests prior to the use of CDR in TAPS. In ASTM foaming tests, the addition of CDR had no significant effect upon the results. In the desalting test, no adverse results were seen on high-temperature emulsion stability. High-temperature heater tests were run, and no heat exchanger fouling tendencies were observed. The coking of crude residue was evaluated with concentrations of CDR of up to 2000 ppm. It is probable that the tank filling procedures, in particular, the filter separator will fully degrade pipeline drag reducer additive by the time it is added to the vehicle tanks (Laura and Tim 2010).

The results of the evaluation of certain parameters such as the number of revolutions, thrust, exhaust temperature, specific fuel consumption, and endoscopic test showed no significant influence of the pipeline drag reducer on the operation of the engine.

TABLE 3.15
Application of Drag Reducers/Flow Improvers Polymer Additives

Problem solved	Location	Liquid	Pipe diameter (inches)	Miles	Throughput MBPD	% D.R. achieved	Approx. flow increase (%)
Increase oil production	Alaska	Crude	16	28	120	38	30
Avoid pump station construction	Alaska	Crude	48	150	1450	22	15
Increase oil production	Offshore Gulf Coast	Crude	6	2	6	16	10
Increase oil production	Far East	Crude	12	14	24	28	20
Increase oil production	Middle East	Crude	40	600	720	22	15
Reduce transport cost	Southwest United States	Crude	16	425	70	19	12
Increase oil production	Far East	Crude	14	96	60	26	18
Increase oil production	Offshore Gulf Coast	Crude	10/12	80	44	38	30
Transport bottleneck	Gulf Coast	Crude	6	11	134	28	20
Reduce transport cost	Midwest United States	Crude	8	31	16	16	10
Seasonal demand	Midwest United States	Diesel	6	51	36	38	30
Seasonal demand	Northeast	Gasoline	8	50	30	28	10
Product mix	Southwest United States	Gasoline	8	50	30	40	30
Winter to summer rate	Midwest United States	No. 2 fuel oil	8	60	25	38	28
Anticipated expanded market	Gulf Coast	Diesel	12	50	95/105	35	20
Proration	Eastern United States	Crude	6	40	15	48	–
Mechanical reliability	North Sea	Crude	16	12	145	38	15
Production surge	Gulf Coast	Crude	12	28	85	39	–
Planning evaluation	Midwest United States	Crude	8	28	35	35	15
Increase allocation	Southwest Asia	Crude	24/36	82	450	21	8
Alternative transport	North Sea	Crude	14/16	30	160	26	18
System bottleneck	Gulf Coast	Crude	28	117	495	23	7

Source: Dean, Hale. June 1984. "Special Report – Slick Way to Increase Capacity". *Pipeline & Gas Journal* 17–19.

The following conclusions are made on the effects of drag reducers on end-use equipment

- The use of drag-reducing additives will result in more carbon deposition in fuels.
- The use of pipeline drag-reducing additives, in particular, CDR 102M, in turbine fuels will result in increased carbon deposition in engine fuel injection nozzles and manifolds where fuel-wetted wall temperatures are 450°F or greater.
- The increased carbon deposition associated with the use of some drag-reducing additives will result in increased maintenance requirements for engines.
- Future engine trends indicate that increased bulk fuel and engine nozzle wetted wall temperatures will make the use of drag-reducing additives such as CDR 102M even more unattractive from a thermal stability standpoint.

3.3.3.3.7 Emulsion Breakers

An emulsion is a mixture of two or more normally immiscible liquids like oil and water. In typical water in oil emulsion, tiny bubbles of water are interspersed within the oil and are stabilized by surface forces. An emulsion breaker destabilizes the bubbles releasing the liquid, which is then forced to form large bubbles, using electrostatics, and settle out by gravity.

Many of the chemical additives discussed here apply only to the oil process, with just a few applicable to gas processing such as corrosion inhibitors and hydrate inhibitors.

3.3.3.4 Dehydration

Dehydration is the most commonly applied measure to protect against internal corrosion in gas pipelines (and in liquid pipelines that contain oil with free water or other electrolytes). Dehydration removes condensation and free water that, if permitted to remain, would allow internal corrosion to occur at points where water droplets precipitate from the gas stream to either form liquid puddles at the bottom of the pipe or adhere to the top of the pipe. Where the gas stream is usually dry, topside corrosion rarely takes place. Complete dehydration is very effective, but because the systems are neither 100% effective nor 100% dependable, there always is the potential to introduce water and other electrolytes into a gas pipeline.

Gas dehydration is used to remove water from the gas. The water content of a gas stream is expressed as:

- Weight of water per volume of gas (mg/Sm³)
- Dew point at a reference pressure (temperature at which the water vapor condenses within the gas)
- A concentration in part per million of water in gas (ppm)

The following determines the water-holding capacity of the gas:

- Higher temperature

- The higher the temperature, the higher the ability of the gas to hold more water
- Lower pressure
- Low pressure increases the water-holding capacity of the gas
- The presence of CO_2 and H_2S in the gas at high pressures
- A gas with these impurities will hold more water at higher pressures, and correction should be made for this when calculating the water content of such a gas stream especially when the gas mixture contains more than 5% H_2S and/or CO_2 at pressures above 4800 kPa.

Typical dehydration strategies for both upstream and downstream section of the gas value chain is:

- Glycol dehydration at gas processing plant (see Figure 3.33)
- Partial dehydration at the wellhead and later additional steps to meet contract specifications
- Chemical injection at the wellhead with later dehydration at the central delivery point
- Full and complete dehydration at each wellhead

3.3.3.4.1   Reason for Dehydrating the Gas

Dehydration is used to control the water dew point. This is important to meet product water dew point specifications and to avoid operational problems.

Water condensation can cause:

- Corrosion: CO_2 and H_2S corrosion in carbon steel pipelines
- Hydrates

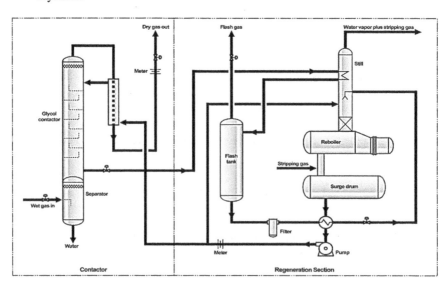

FIGURE 3.33   Typical glycol dehydration unit (GDU).

3.3.3.4.2 Common Gas Dehydration Methods

Main technologies used for gas dehydration are:

a. Glycol dehydration (physical absorption) (MEG, DEG, TEG, TREG)
b. Adsorption on the solid bed (e.g. molecular sieves)
c. LTS (low-temperature separators) with glycol injection system (TEG)

3.3.3.4.2.1 Glycol Dehydration

Glycol dehydration is a liquid desiccant system for the removal of water from natural gas and natural gas liquids (NGLs). It is the most common and economical means of water removal from these streams. Glycols typically seen in the industry include tri-ethylene glycol (TEG), diethylene glycol (DEG), mono-ethylene glycol, and tetra-ethylene glycol (TREG). TEG is the most commonly used glycol in the industry.

The process is carried out by the absorption of the water vapors in a glycol solution in a contactor tower. Highly concentrated glycol solutions (TEGs) are used to physically absorb the water from the gas, and the glycol solution is then regenerated so it can be used again.

A typical dehydration unit is presented in Figure 3.33. The design can vary from plant to plant.

3.3.3.4.2.2 Adsorption on Solid Bed (e.g. Molecular Sieves)

Separation processes by adsorption use a solid bed material with a large specific surface area. There are several solid desiccants that possess the physical characteristic to adsorb water from natural gas (see Figure 3.34).

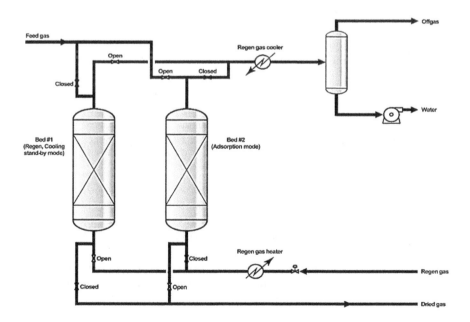

FIGURE 3.34 Typical gas dehydration, by adsorption process.

The solid desiccants for commercial use are divided into three categories:

- Gels (alumina or silica): silica gel is generally used for dehydration and i-C5+ recovery from the natural gas stream. When silica gel is used for dehydration, the achieved water dew point is approximately −60°C. Silica gel is the easiest desiccant to regenerate, and although absorbing heavy hydrocarbons they are more easily removed in regeneration.
- Alumina (activated aluminum oxide): alumina as solid desiccant reaches water dew point down to −73°C.
- Molecular sieves (alumina-silicates): molecular sieves have the highest water capacity and can get down to a water dew point of −100°C. Because they are selective absorbers, heavy hydrocarbons are not absorbed; however, they require the most energy to regenerate the desiccant.

The adsorption on the solid bed process requires two (or more) vessels, with one of the vessels removing water while the other is being regenerated and cooled.

Usually one adsorption cycle can last from 8 hours to 24 hours.

In general, four methods are commercially available in order to carry out the regeneration of saturated desiccant material:

- Temperature swing adsorption (TSA)
- Pressure swing adsorption (PSA)
- Inert purge stripping

Evaluation of the type of solid desiccant system required may be based on the following factors:

1. The dew point required: molecular sieves give the lowest dew point, which may be required prior to cryogenic processing.
2. The size of drying beds: on an offshore platform the available space will be limited. The absorption capacity of each desiccant at the dryer operating conditions must be studied.
3. The gas composition: larger hydrocarbon molecules are absorbed by some desiccants. The larger molecules will be released during regeneration and form part of the fuel gas stream. These larger molecules, which may be valuable products, cannot then be recovered in the processing facilities.
4. Cycle period: the drying period prior to regeneration governs the size of the driers. For a short drying period the size of the driers may be smaller. However, with more rapid cycling there is more likelihood of mechanical failure of the changeover systems. Optimum drying periods are of the order of 24–48 hours.
5. Regeneration requirements: for higher-regeneration temperature requirements the larger the regeneration heating facilities must be. Again space limitations must be considered.

3.3.3.4.2.3 Low-Temperature Separator (LTS) with Glycol Injection System A low-temperature separator with glycol injection system is used if the water or HC dew point cannot be met and also to inhibit hydrate formation. It is typically used when it is not cost effective to install a full glycol dehydration unit (GDU) or gas dehydration unit or a solid adsorption unit.

The gas is cooled to condense both water and hydrocarbons (NGL). Cooling is performed using mechanical refrigeration, J–T effect, or Turbo-expander. The water and hydrocarbon dew point are determined by the operating temperature of the cold separator. The glycol injection prevents the formation of hydrates, which can block the equipment. This process is illustrated in Figure 3.35.

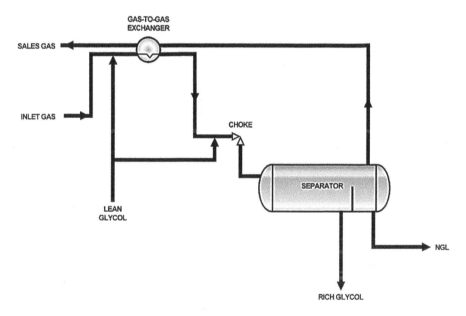

FIGURE 3.35 Gas dehydration module.

3.3.3.5 Cleaning Pigs

Pig is an abbreviation for the pipeline inspection gauge. Pigging is an in-line inspection (ILI) method in which devices, known as "pigs", are inserted into pipelines to clean and/or inspect the pipeline. A pig is a device inserted into a pipeline to perform various maintenance operations. These operations include cleaning the pipeline, inspection to provide information on the condition of the pipeline, and the extent and location of any problem such as corrosion and isolation of a pipeline by plugging it. A pig is launched from a pig-launching unit and is driven by the product fluid. It is received at the other end by a pig receiver device.

The main benefits of pigging a pipeline are to clear the line of debris and also to gather information about the pipeline. The user then uses this information to map out defects to aid repair crews in maintenance.

Cleaning pigs can effectively direct both liquids and corrosive solids to pig traps for removal from the pipeline. Routine maintenance pigging will direct any liquid pools away from low points and, if performed properly, out of the entire pipeline. Cleaning pigs will also move the solids and remove them from the pipeline through the pig trap at the end of the pipeline (Figure 3.36).

FIGURE 3.36   Cleaning pig (Source: https://smartpigs.net/pigging-products.html).

These pigs are specially made to remove sedimentation and buildup that can impede the flow of materials.

There are two ways of cleaning a pipeline

a. Mechanical cleaning

   The generally approved and accepted practice among pipeline operators to remove deposits in a pipeline is by *mechanical pigging*. The pig is repetitively sent through the pipeline to swap deposits from that pipeline until hardly any deposit can be found in the pig receiving station. It is, however, hard to determine if this implies that the pipeline is clean.

b. Advanced chemical cleaning

   'Advanced chemical cleaning" is rapidly becoming an industry standard. Chemical cleaning in conjunction with the use of mechanical pigs removes a greater volume of debris with fewer runs. Chemical cleaning, by definition, means the use of liquid cleaners mixed in a diluent (water, diesel, methanol, Iso-propyl-alcohol and the like) to form a cleaning solution that can be pushed through a pipeline by pigs. For example, the removal of hydrocarbon hydrates from a pipeline system involves running a batch of either methanol or glycol through the pipeline, driven by a pig. The size of the batch should be sufficient to depress the hydrate formation temperatures of the maximum anticipated hydrate deposition, to below the pipeline temperature, as well as allowing for liquid that will be left on the pipe wall and not pushed forward by the pig. Removal of the hydrates from raw natural gas pipelines, which may not have pigging facilities, is usually achieved by increasing the dosage rate of hydrate point depressant. For a short period, shock dosing, at a rate of approximately five times the normal injection rate, should be undertaken.

Selection of designed pipeline cleaners should be based upon the following features
- A neutral pH
- Deposit permeating and penetrating capabilities
- Original design parameters of the cleaner and its case histories
- Health Safety and Environment (HSE) awareness

3.3.3.6 Buffering

A buffering agent is a weak acid or base used to maintain the acidity (pH) of a solution near a chosen value after the addition of another acid or base. That is, the function of a buffering agent is to prevent a rapid change in pH when acids or bases are added to the solution.

Buffering agents change the chemical composition of fluids that remain in the pipeline and hence could be utilized to prevent internal corrosion. A buffering agent, such as a mild or dilute alkaline mixture, can significantly reduce the corrosivity of any standing liquid, predominantly by raising its pH value above 7 (neutral), so that it turns from acidic to alkaline. Alkaline liquids cause virtually no harm to steel. In general, buffering is not very effective because it is difficult to cover the entire pipe surface.

The use of buffers or other pH-altering chemicals also can modify the environment and possibly eliminate the growth of the bacteria that cause microbial corrosion (Baker 2008) (Table 3.16).

3.3.3.7 Corrosion Monitoring Techniques

Table 3.17 describes the most common techniques for monitoring corrosion and operating conditions associated with internal corrosion in gas pipelines.

3.3.3.8 Corrosion Inspection Techniques

Table 3.18 describes common techniques that should be considered for the detection of internal corrosion in gas pipelines. Note: Due to localized corrosion being the prevalent failure mechanism in sour gas pipelines, hydrotesting alone may not be adequate to prove pipeline integrity.

TABLE 3.16

Practices for Mitigating Internal Corrosion During Operation

Technique	Description	Comments
Production monitoring	Monitor fluid (gas) temperature	The temperature of the soil, as well as the temperature of the pipe, may create favorable conditions for an attack on pipeline materials. Liquid and gas lines have slightly different operating temperature characteristics, but both are still susceptible. For example, with gas pipelines both the pipe and the surrounding ground can vary from a high of 40°C upon leaving the compressor station down to 5°C at distances from the station.
Cleaning pigging	Periodic pigging of pipeline segments to remove liquids, solids, and debris	• Pigging is one of the most effective methods of internal corrosion control • Can be an effective method of cleaning pipelines and reducing the potential for bacteria colonization and under-deposit corrosion • Selection of the pig type and sizing is important • Requires facilities for launching and receiving pigs
Batch corrosion inhibition (chemical treating)	• Periodic application of a batch corrosion inhibitor to provide a protective barrier on the inside surface of the pipe • Initial batch treatment of the pipeline is critical at pipeline commissioning, after new pipeline construction, repairs, or suspension • Batching is required after any activity that will disrupt the protective films (inspection, line repairs, workovers, etc.)	• Provides a barrier between corrosive elements and pipe surface • The application procedure is important in determining effectiveness (e.g. volume and type of chemical, diluent used, contact time, and application interval) • Effectiveness may be reduced on existing pitting, in particular, deep and narrow morphology • Should be applied between two pigs to effectively clean and lay down inhibitor on the pipe • Should be used in conjunction with pigging to remove liquids and solids (i.e. the inhibitor must be applied to clean pipe to be the most effective) • Some corrosion inhibitors may not be effective in top-of-the-line corrosion • Should be used in conjunction with pigging to remove liquids and solids (i.e. the inhibitor must be applied to clean pipe to be the most effective)

(Continued)

TABLE 3.16 (CONTINUED)
Practices for Mitigating Internal Corrosion During Operation

Technique	Description	Comments
	Note: Large-diameter lines may require special design and/or procedures to ensure batch slug remains intact. Batch programs have numerous variables (including people, chemical, and application) and need to be properly managed to ensure effective implementation and performance monitoring	
Continuous corrosion inhibition (chemical treating)	Continuous injection of a corrosion inhibitor to reduce the corrosivity of the transported fluids or provide a barrier film	• Corrosion inhibitor may be less effective at contacting full pipe surface especially in a dirty system batch may be more effective • Program design is important (e.g. product selection, performance criteria, production characteristics) • Can help with top-of-the-line corrosion mitigation
Biocide chemical treating	Periodic application of a biocide to kill bacteria in the pipeline system	• Assist in controlling bacterial growth or killing bacteria in systems known to contain bacteria • Use in conjunction with pigging (to clean the line) will enhance effectiveness • Batch application typically most effective (e.g. application downhole leads to the ongoing treatment of produced fluids flowing into the pipeline) • The use of improperly selected biocides can create a foam that can be a serious operational issue
Oxygen control	• Use gas blanketing and oxygen (O_2) scavengers • Avoid purging test equipment into the pipeline • Optimize methanol injection and/or use inhibited methanol • Batch-treat pipelines following line repairs, inspections, and hydrotesting	• Oxygen ingress will accelerate the corrosion potential (can also create sulfur compounds)

TABLE 3.17
Internal Corrosion Monitoring Technique

Technique	Description	Comments
Gas and oil analysis	Ongoing monitoring of gas composition for H_2S and CO_2 content. If present, the analysis of liquid hydrocarbon properties is useful	• Acid gas content must be understood and should be periodically re-assessed • Trend of reservoir souring should be monitored
Water analysis	Ongoing monitoring of water for chlorides, dissolved metals, suspended solids, and chemical residuals	• Changes in water chemistry will influence the corrosion potential • Trends in the concentration of dissolved metals (e.g. Fe, Mn) can indicate changes in corrosion activity (monitoring of iron–manganese ratio may not be as effective in the H_2S system) • Chemical residuals can be used to confirm the level of application and changes in water production • Sampling location and proper procedures are critical for accurate results
Production monitoring	Ongoing monitoring of production conditions such as pressure, temperature, and flow rates	• Changes in operating conditions will influence the corrosion potential • Production information can be used to assess corrosion susceptibility based on fluid velocity and corrosivity
Mitigation program compliance	Ongoing monitoring of mitigation program implementation, execution, and documentation	• Chemical pump reliability and inhibitor inventory control are critical where the mitigation program includes continuous chemical injection • Corrosion mitigation program must be properly implemented and maintained to be effective • Impact of any non-compliance to the mitigation program must be evaluated to assess the effect on corrosion
Corrosion coupons	Used to indicate general and pitting corrosion susceptibility and mitigation program effectiveness	• Trends in coupon data can indicate changes in corrosion activity • Coupons should be used in conjunction with other monitoring and inspection techniques • Coupon type, placement, and data interpretation are critical to the successful application of this method

(Continued)

TABLE 3.17 (CONTINUED)
Internal Corrosion Monitoring Technique

Technique	Description	Comments
Bio-spools	Used to monitor for bacteria presence and biocide effectiveness	• Bio-spool placement and data interpretation are critical to the successful application of these methods • Bio-spools should be used in conjunction with other monitoring and inspection techniques • Solids pigged out of pipelines (pig yields) can be tested for sessile bacteria levels • The presence of bacteria on surfaces is considered a better way to quantify the type and numbers present in the system
Electrochemical monitoring	There are a variety of methods available such as electrochemical noise, linear polarization, electrical resistance, hydrogen foils/probes, and field signature method	• Device selection, placement, and data interpretation are critical to the successful application of these methods • Continuous or intermittent data collection methods are used • Electrochemical monitoring should be used in conjunction with other monitoring and inspection techniques

3.3.4 MITIGATION OF INTERNAL CORROSION IN CARBON STEEL OIL-EFFLUENT PIPELINE SYSTEMS

This chapter intends to assist upstream oil and natural gas producers in recognizing the conditions that contribute to internal pipeline corrosion and identify measures to reduce the likelihood of internal corrosion incidents. Specifically, this document addresses design, maintenance, and operating considerations for the mitigation of internal corrosion in oil-effluent pipeline systems.

3.3.4.1 Corrosion Mechanisms and Mitigation

Pitting corrosion along the bottom of the pipeline is the primary corrosion mechanism leading to failures in oil-effluent pipelines. The common features of this mechanism are:

- Presence of water containing any of the following: CO_2, H_2S, chlorides, bacteria, O_2, or solids
- Pipelines carrying higher levels of free-water production (high water/oil ratio or water cut)
- Presence of liquid traps where water and solids can accumulate or are left (e.g. inactive pipelines) (Figure 3.37)

TABLE 3.18

Internal Corrosion Inspection Technique (CAPP 2018)

Options	Technique	Comments
In-line inspection	• Magnetic flux leakage, ultrasonic, and eddy current tools are available. MFL is the most commonly used technique. • UT and eddy current tools are also available	• Effective method to accurately determine the location and severity of corrosion • In-line inspection can find internal and external corrosion defects • The tools are available as self-contained or tethered • The pipeline must be designed or modified to accommodate in-line inspection • To run a tethered tool inspection, it is often necessary to dig bell holes and cut the pipeline
Non-destructive examination	Ultrasonic inspection, radiography, or other NDE methods can be used to measure metal loss in a localized area	• An evaluation must be done to determine potential corrosion sites prior to conducting NDE • See NACE SP0110 Wet Gas Internal Corrosion Direct Assessment Methodology for Pipelines or NACE SP0206 Internal Corrosion Direct Assessment Methodology for Pipelines Carrying Normally Dry Gas • The use of multi-film radiography is an effective screening tool prior to using ultrasonic testing • NDE is commonly used to verify in-line inspection results, corrosion at excavation sites, and aboveground piping • Practical limitations of NDE methods and the factors affecting accuracy must be understood
Video camera/ boroscopes	Visual inspection tool to locate internal corrosion	• Used to locate and determine the presence of corrosion damage, but it is difficult to determine severity • Technique may be limited to short inspection distances • Cannot directly measure the depth of corrosion pits
Destructive examination	Physical cutout of sections from the pipeline	• Consideration should be given to locations where specific failure modes are most likely to occur • See NACE SP0110 Wet Gas Internal Corrosion Direct Assessment Methodology for Pipelines or NACE SP0206 Internal • Corrosion Direct Assessment Methodology for Pipelines Carrying Normally Dry Gas

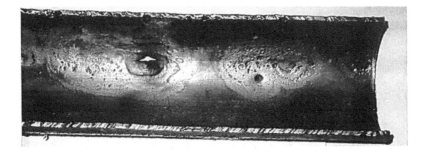

FIGURE 3.37 Example of internal corrosion in an oil-effluent pipeline (Source: CAPP. 2018. *Mitigation of internal corrosion in carbon steel oil effluent pipeline systems.* Calgary, Alberta: Canadian Association of Petroleum Producers (CAPP)).

Table 3.19 describes the most common contributors, causes, and effects of internal corrosion in oil-effluent pipelines. They also contain common mitigation measures to reduce pipeline corrosion. They apply to internally bare carbon steel pipeline systems and coated or lined pipelines where deterioration or damage has allowed water contact with the steel substrate.

3.3.4.2 Recommended Practices

Recommended practices for the mitigation of internal corrosion during the design and construction of oil-effluent pipelines are described in Table 3.20.

Table 3.21 describes the recommended practices for the mitigation of internal corrosion during operations of oil-effluent pipelines.

3.3.4.3 Corrosion Mitigation Techniques

Table 3.22 describes common techniques to be considered for the mitigation of internal corrosion in oil-effluent pipelines.

3.3.4.4 Corrosion Monitoring Techniques

Table 3.23 describes the most common techniques for monitoring corrosion and operating conditions associated with internal corrosion in oil-effluent pipelines.

3.3.4.5 Corrosion Inspection Techniques

Table 3.24 describes common techniques to be considered for the detection of internal corrosion in oil-effluent pipelines.

3.3.4.6 Repair and Rehabilitation Techniques

Table 3.25 describes common techniques for repair and rehabilitation of pipelines damaged by internal oil-effluent corrosion.

Prior to the repair or rehabilitation of a pipeline, the appropriate codes and guidelines should be consulted, including:

- CSA Z662, Clause 10 Including Temporary and Permanent Repair Methods
- CSA Z662 Clause 13 Reinforced composite, thermoplastic-lined, and polyethylene pipelines

TABLE 3.19

Contributing Factors and Prevention of Internal Oil-Effluent Corrosion – Mechanisms (CAPP 2018)

Contributor	Cause/source	Effect	Mitigation
Water holdup	• Low velocity and poor pigging practices allow water to stagnate in the pipelines • Deadlegs or inactive service	• Water acts as the electrolyte for the corrosion reaction • Chlorides increase the conductivity of water and may increase the localized pitting rate • In some cases, pitting corrosion may not occur if chloride ion concentration is very high ($>10\%$)	• Install pigging facilities and maintain an effective pigging program • Control corrosion through effective corrosion inhibition practices • Remove stagnant deadlegs • Effectively pig lines as soon as the wells become inactive
Carbon dioxide	• Produced with gas from the reservoir • CO_2 flooding	• CO_2 dissolves in water to form carbonic acid • Corrosion rates increase with increasing CO_2 partial pressures	• Effective pigging and inhibition
Hydrogen sulfide (H_2S)	• Produced with gas from the reservoir • Generated by sulfate-reducing bacteria	• Hydrogen sulfide can form protective iron sulfide scales to reduce general corrosion rates • Localized breakdown of iron sulfide scales triggers pitting corrosion attack	• Effective pigging and inhibition programs • See "Bacteria"
Bacteria	• Contaminated drilling and completion fluids • Contaminated production equipment • Produced fluids from the reservoir (contaminated)	• Acid-producing and sulfate-reducing bacteria can lead to localized pitting attack • Solid deposits provide an environment for the growth of bacteria	• Treat with corrosion inhibitors and biocides • Effective pigging program • Treat any introduced fresh water with biocide
Critical velocity	• Critical velocity is reached when there is insufficient flow to sweep the pipeline of water and solids	• Water accumulation and solid deposition (inorganic sands, mineral scales, corrosion products, elemental sulfur, etc.) accelerate corrosion	• Design pipeline to exceed critical velocity • Establish operating targets based on critical gas and liquid velocity to trigger appropriate mitigation requirements, e.g. pigging and batch inhibition

(Continued)

TABLE 3.19 (CONTINUED)
Contributing Factors and Prevention of Internal Oil-Effluent Corrosion – Mechanisms (CAPP 2018)

Contributor	Cause/source	Effect	Mitigation
Solids deposition	• Mainly produced from the formation; can include wax, asphaltenes, scales, and sands • May originate from drilling fluids, workover fluids, and scaling waters • May include corrosion products from upstream equipment • Insufficient velocities and poor pigging practices	• Can contribute to under-deposit corrosion • Scaling can interfere with corrosion monitoring and inhibition • Solids will reduce the corrosion-inhibiting efficiency	• Install pigging facilities and maintain an effective pigging program • Use wellsite separators to tank and truck water during workover and completion activities to minimize the effects on the pipeline • Scale suppression
Drilling and completion fluids	• Introduction of bacteria • Introduction of spent acids and kill fluids • Introduction of solids	• Accelerated corrosion • Lower pH • Higher chloride concentration, which can accelerate corrosion and reduce the corrosion inhibitor dispersability	• Produce wells to wellsite separator, tanking, and trucking water until drilling and completion fluids and solids are recovered • Supplemental pigging and batch inhibition of pipelines before and after workover activities
Detrimental operating practices	• Ineffective pigging • Ineffective inhibition • Inadequate pipeline suspension activities • Commingling of incompatible produced fluids	• Accelerated corrosion	• Design pipelines to allow for effective shut-in and isolation • Develop and implement proper suspension procedures, including pigging and batch inhibition • Test for fluid incompatibilities

(Continued)

TABLE 3.19 (CONTINUED)
Contributing Factors and Prevention of Internal Oil-Effluent Corrosion – Mechanisms (CAPP 2018)

Contributor	Cause/source	Effect	Mitigation
Management of change	• Change in production characteristics or operating practices • Well re-completions and work overs • Lack of system operating history and practices • Changing personnel and system ownership	• Unmanaged change may result in accelerated corrosion	• Implement an effective MOC process as part of the IMP • Maintain the integrity of pipeline operation and maintenance history and records • Re-assess corrosivity on a periodic basis
Oxygen ingress	• Completion fluids or other fluids saturated with O_2 • Tanks and other wellsite equipment (e.g. pump jack compressors) • Methanol injection	• Accelerated corrosion • Can precipitate elemental sulfur if service is sour • Corrosion inhibitors may be ineffective if O2 is present	• Scavenge O_2 or deaerate
• Stray current corrosion	• Conductive bridge forming across insulating kits • High chloride service and high electrode potential differential	• Aggressive flange face corrosion	• Move insulating kit to a vertical position • Use thicker isolating gasket or use an electrical isolating joint • Short or remove insulating kit if the CP system can handle the added load • Insert a resistance bond or modify the CP system to reduce the potential differential • Internally coat flange and short spool of piping with high dielectric coating, either on both sides of the flange or at least on the side that is protected with CP

TABLE 3.20

Recommended Practices – Design and Construction (CAPP 2018)

Element	Recommended Practice	Benefit	Comments
Materials of construction	• Use normalized electric resistance welding (ERW) line pipe that meets the requirements of CSA Z245.1 Steel Pipe • Use CSA Z245.1 Sour Service Steel Pipe for sour oil-effluent pipelines • Consider the use of corrosion-resistant materials such as high-density polyethylene (HDPE) or fiber-reinforced composite materials as per CSA-Z662, Clause 13 Plastic Pipelines	• Normalized ERW prevents preferential corrosion of the weld zone • Reduces chances of sulfide stress cracking failures • Non-metallic materials are corrosion resistant	• EW seams should be placed on the top half of the pipe to minimize preferential corrosion • Non-metallic materials may be used as a liner or a free-standing pipeline depending on the service conditions. Be aware that internally bare steel risers would be susceptible to corrosion
Pipeline isolation	• Install valves that allow for the effective isolation of pipeline segments from the rest of the system • Install the valves as close as possible to the tie-in point • Install blinds for the effective isolation of inactive pipeline segments	• Allows for more effective suspension and discontinuation of pipeline segments	• Removes potential "deadlegs" from the gathering system • Be aware of creating "deadlegs" between isolation valve and mainline at tie-in locations, which can be mitigated through tee tie-ins at 12 o'clock or aboveground riser tie-ins • Develop shut-in guidelines for the timing of requiring steps to isolate and lay-up pipelines in each system

(Continued)

TABLE 3.20 (CONTINUED)
Recommended Practices – Design and Construction (CAPP 2018)

Element	Recommended Practice	Benefit	Comments
Pipeline sizing	• Design pipeline system to ideally maintain flow above critical velocity	• Using optimal line size where possible, increases velocity and reduces water stagnation and solids accumulation	• Consider future operating conditions such as changes in well deliverability • Consider the future corrosion mitigation cost of oversized pipelines operating under the critical velocity • Consider the impact of crossovers, line loops, and flow direction changes
Pigging capability	• Install or provide provisions for pig launching and receiving capabilities • Use consistent line diameter and wall thickness • Use piggable valves, flanges, and fittings	• Pigging is one of the most effective methods of internal corrosion control • Pigging can improve the effectiveness of corrosion inhibitors	• Multi-disk/cup pigs have been found to be more effective than ball or foam type pigs • Receivers and launchers can be permanent or mobile • Use pigs that are properly oversized, undamaged, and not excessively worn • Pigging may be ineffective when there are large differences in wall thickness or line sizes
Inspection capability	• Install or provide capability for in-line inspection tool launching and receiving • Use consistent line diameter and wall thickness • Use piggable valves, flanges, and fittings	• Internal inspection using in-line inspection is the most effective method for confirming overall pipeline integrity • Proper design allows for pipeline inspection without costly modifications or downtime	• Consideration should be given to the design of bends, tees, and risers to allow for navigation by the inspection devices

TABLE 3.21

Recommended Practices – Operations (CAPP 2018)

Element	Recommended practice	Benefit	Comments
Completion and workover practices	• Produce wells to wellsite separation, tanking, and trucking of water until completion and workover fluids and solids are recovered	• Removal of completion and/or workover fluids reduces the potential for corrosion	• Supplemental pigging and batch inhibition of pipelines may be required before and after workover activities
Corrosion assessment	• Evaluate operating conditions (temperature, pressure, well effluent and volumes) and prepare a corrosion mitigation program • Develop and communicate corrosion assessment, operating parameters and the mitigation program with all key stakeholders including field operations and maintenance personnel • Re-assess corrosivity on a periodic basis and subsequent to a line failure	• Effective corrosion management comes from understanding and documenting design and operating parameters	• Refer to CSA Z662 Clause 9 – Corrosion Control • Define acceptable operating ranges consistent with the mitigation program (See CSA Z662 Clause 10) • Consider the effects of oxygen, methanol, bacteria and solids • Consider supplemental requirements for handling completion and workover fluid backflow
Corrosion inhibition and monitoring	• Develop and communicate the corrosion inhibition and monitoring program with all key stakeholders including field operations and maintenance personnel • Develop suspension and lay-up procedures for inactive pipelines	• Allows for an effective corrosion mitigation program	• Refer to Section 3.3.4.3 for Corrosion Mitigation Techniques • Refer to Section 3.3.4.4 for Corrosion Monitoring Techniques • Refer to CSA Z662 Clause 9 – Corrosion Control • Number and location of monitoring devices is dependent on the predicted corrosivity of the system • Process sampling for monitoring such as bacteria, Cl–, pH, Fe, Mn, and solids • Consider provisions for chemical injection, monitoring devices, and sampling points

(Continued)

TABLE 3.21 (CONTINUED)
Recommended Practices – Operations (CAPP 2018)

Element	Recommended practice	Benefit	Comments
Inspection program	• Develop an inspection program or strategy • Communicate the inspection program to field operations and maintenance personnel	• Creates greater "buy in" and awareness of corrosion mitigation program • Provides assurance that the corrosion mitigation program is effective	• Refer to Section 3.3.4.5 for Corrosion Inspection Techniques • Refer to CSA Z662 Clause 9 – Corrosion Control
Failure investigation	• Recovery of an undisturbed sample of the damaged pipeline • Conduct thorough failure analysis • Use the results of the failure analysis to re-assess the corrosion mitigation program	• Improved understanding of corrosion mechanisms detected during inspections or as a result of a failure • Allows for corrosion mitigation program adjustments in response to inspection results	• Adjust the corrosion mitigation program based on the results of the failure analysis • Some onsite sampling may be required during sample removal (e.g. bacteria testing and elemental sulfur)
Repair and rehabilitation	• Inspect to determine extent and severity of damage prior to carrying out any repair or rehabilitation • Based on inspection results, use CSA Z662 Clause 10 to determine the extent and type of repair required • Implement or make modifications to integrity management program after repairs and failure investigations so that other pipelines with similar conditions are inspected and mitigation programs revised as required	• Prevents multiple failures on the same pipeline • Prevents reoccurrence of problem in other like pipelines in the system	• Refer to Section 3.3.4.5 for Corrosion Inspection Techniques • Refer to Section 3.3.4.6 for Repair and Rehabilitation Techniques • Refer to CSA Z662 Clause 10 for repair requirements
Leak detection	• Integrate a leak detection strategy	• Permits the detection of leaks	• The technique utilized depends on access and ground conditions
Management of change	• Implement an effective MOC process • Maintain pipeline operation and maintenance records	• Ensures that change does not impact the integrity of the pipeline system • Understand and document design and operating parameters	• Unmanaged change may result in accelerated corrosion, using inappropriate mitigation strategy for the conditions (outside the operating range)

TABLE 3.22

Corrosion Mitigation Techniques (CAPP 2018)

Technique	Description	Comments
Pigging	• Periodic pigging of pipeline segments to remove liquids, solids, and debris	• Pigging is one of the most effective methods of internal corrosion control • Can be an effective method of cleaning pipelines and reducing the potential for bacteria colonization and under-deposit corrosion • Selection of pig type and sizing is important • Requires facilities for launching and receiving pigs
Batch Corrosion Inhibitor Chemical Treating	• Periodic application of a batch corrosion inhibitor to provide a protective barrier on the inside surface of the pipe	• Provides a barrier between corrosive elements and pipe surface • Application procedure is important in determining effectiveness (e.g. volume and type of chemical, the diluent used, contact time, and application interval) • Effectiveness may be reduced on existing pitting, in particular, deep and narrow morphology • Should be applied between two pigs to effectively clean and lay down inhibitor on the pipe • Should be used in conjunction with pigging to remove liquids and solids (i.e. the inhibitor must be applied to clean pipe to be the most effective)
Continuous Corrosion Inhibitor Chemical Treating	• Continuous injection of a corrosion inhibitor to reduce the corrosivity of the transported fluids or provide a barrier film	• Less common technique due to the high cost to treat high-volume-water-producing wells • Corrosion inhibitor may be less effective at contacting full pipe surface especially in a dirty system, batch may be more effective • Chemical pump reliability is important in determining effectiveness
Biocide Chemical Treating	• Periodic application of a biocide to kill bacteria in the pipeline system	• Assists in controlling bacterial growth • Use in conjunction with pigging (to clean the line) will enhance the effectiveness • Batch application typically most effective (e.g. application downhole leads to ongoing treatment of produced fluids flowing into the pipeline) • The use of improperly selected biocides can create a foam that can be a serious operational issue

TABLE 3.23
Corrosion Monitoring Techniques (CAPP 2018)

Technique	Description	Comments
Gas and Oil Analysis	• Ongoing monitoring of gas composition for H2S and CO_2 content. The analysis of liquid hydrocarbon properties is useful	• Acid gas content must be understood and should be periodically re-assessed.
Water Analysis	• Ongoing monitoring of water for chlorides, pH dissolved metal ions, bacteria, total dissolved solids (TDS), suspended solids and chemical residuals	• Changes in water chemistry will influence the corrosion potential. • Trends in dissolved metal ion concentration (e.g. Fe, Mn) can indicate changes in corrosion activity. • Chemical residuals can be used to confirm the proper concentration of corrosion inhibitors. • Sampling location and proper procedures are critical for accurate results.
Production Monitoring	• Ongoing monitoring of production conditions such as pressure, temperature and flow rates • Ongoing monitoring of changes in water–oil ratio (water cut)	• Changes in operating conditions will influence the corrosion potential. Production information can be used to assess corrosion susceptibility based on fluid velocity and corrosivity. • Increasing water-cut tends to have a higher likelihood of internal corrosion. Depending on a number of factors, typically a higher water-cut will lead to water-wet steel surface. Some liquid hydrocarbons will have better protective properties than others.
Mitigation Program Compliance	• Ongoing monitoring of mitigation program implementation and execution	• Chemical pump reliability, injection rate targets and inhibitor inventory control are critical where mitigation program includes continuous chemical injection. • The corrosion mitigation program must be properly implemented to be effective. • The impact of any non-compliance to the mitigation program must be evaluated to assess the effect on corrosion.

(Continued)

TABLE 3.23 (CONTINUED)
Corrosion Monitoring Techniques (CAPP 2018)

Technique	Description	Comments
Corrosion Coupons	• Used to indicate general corrosion rates, pitting susceptibility, and mitigation program effectiveness	• Coupon type, placement, and data interpretation are critical to the successful application of this method. • Coupons should be used in conjunction with other monitoring and inspection techniques.
Bio-spools	• Used to monitor for bacteria presence and mitigation program effectiveness	• Bio-spool placement and data interpretation are critical to the successful application of these methods. • Bio-spools should be used in conjunction with other monitoring and inspection techniques. • Solids pigged out of pipelines (pig yields) can be tested for bacteria levels. • Consider following NACE TM0212. • The presence of bacteria on pipeline internal surfaces is considered a better way to quantify type and numbers present in the system.
Electrochemical Monitoring	• There are a variety of methods available such as electrochemical noise, linear polarization, electrical resistance, and field signature method	• The device selection, placement, and data interpretation are critical to the successful application of these methods. • Continuous or intermittent data collection methods are used. • Electrochemical monitoring should be used in conjunction with other monitoring and inspection techniques.

TABLE 3.24

Corrosion Inspection Techniques (CAPP 2018)

Options	Technique	Comments
In-line Inspection	• Magnetic flux leakage is the most common technique	• In-line inspection data should be verified using other methods • Effective method to determine location and severity of corrosion along the steel pipelines • In-line inspection can detect both internal and external corrosion wall loss as well as other types of imperfections • The pipeline must be designed or modified to accommodate in-line inspection • The tools are available as free swimming or tethered • To run a tethered tool inspection, it is often necessary to dig bell holes and cut the pipeline
Non-Destructive Examination (NDE)	• Ultrasonic inspection, radiography or other NDE methods can be used to measure metal loss in a localized area	• Evaluation must be done to determine potential corrosion sites prior to conducting NDE (see NACE SP0116 Multiphase Flow Internal Corrosion Direct Assessment Methodology for Pipelines) • NDE is commonly used to verify in-line inspection results and corrosion at excavation sites. NDE methods may be used to measure corrosion pit growth at excavation sites; however, the practical limitations of NDE methods and the factors affecting accuracy must be understood. • The use of radiography is an effective screening tool prior to using ultrasonic testing • Corrosion rates can be determined by performing periodic NDE measurements at the same locations
Video Camera/ Boroscope	• Used as a visual inspection tool to locate internal corrosion	• Can be used to determine the presence of corrosion damage, but it is difficult to determine severity • This technique may be limited to short inspection distances
Destructive Examination	• Physical cut out and examination of sections from the pipeline	• Consideration should be given to locations where specific failure modes are most likely to occur. (see NACE SP0116 Multiphase Flow Internal Corrosion Direct Assessment Methodology for Pipelines)

TABLE 3.25

Repair and Rehabilitation Techniques (CAPP 2018)

Technique	Description	Comments
Pipe Section Replacements	• Remove damaged section(s) and replace	• When determining the quantity of pipe to replace, consider the extent of the corrosion and as well as the extent and severity of damage or degradation of any internal coatings or linings along with the condition of the remaining pipeline • Impact on pigging capabilities must be considered (use same pipe diameter and similar wall thickness) • The replaced pipe section should be coated with corrosion inhibitor prior to commissioning or coated with an internal coating compatible with the existing pipeline
Repair Sleeves	• Reinforcement and pressure-containing sleeves may be acceptable for temporary or permanent repairs of internal corrosion as per the limitations stated in CSA Z662	• For internal corrosion it may be possible in some circumstances for the damaged section to remain in the pipeline as per the requirements in CSA Z662 Clause 10 along with proper corrosion control practices to prevent further deterioration • Different repair sleeves are available including composite, weld-on and bolt-on types. The sleeves must meet the requirements of CSA Z662 Clause 10
Thermoplastic Liners	• A polymer liner is inserted in the steel pipeline • The steel pipe must provide the pressure containment capability	• A variety of materials are available with different temperature and chemical resistance capabilities • Impact on pigging capabilities must be considered • Polymer liners may eliminate the need for internal corrosion mitigation, corrosion monitoring and inspection • Reduction of inhibition programs may impact the integrity of connecting headers and facilities constructed from internally bare carbon steel
Composite or Plastic Pipeline	• Free-standing composite or plastic pipe can be either plowed-in for new lines, or pulled through old pipelines • This pipe must be designed to provide full pressure containment	• A variety of materials are available with different temperature and chemical resistance capabilities • Free-standing plastic pipelines are typically limited to low-pressure service • Impact on pigging capabilities and pig selection must be considered • Composite or plastic pipelines may eliminate the need for internal corrosion mitigation, corrosion monitoring and inspection • Reduction of inhibition programs may impact the integrity of connecting headers and facilities constructed from internally bare carbon steel

(Continued)

TABLE 3.25 (CONTINUED)
Repair and Rehabilitation Techniques (CAPP 2018)

Technique	Description	Comments
Entire Pipeline Replacement	• Using internally coated steel pipeline systems with an engineered joining system should also be considered • The alteration or replacement of the pipeline allows for proper mitigation and operating practices to be implemented	• Should be pig and inspection tool compatible (required for sour systems per CSA Z662 Clause 16) • Refer to Section 3.3.4.2 "Recommended Practices" in this document for details • Ensure that when replacements in kind occur, the replacement of the pipeline allows for proper mitigation and operating practices to be implemented

3.4 AC CORROSION MITIGATION

AC corrosion refers to corrosion initiated and propagating under the influence of alternating current. AC voltage on a buried or immersed pipeline is the driving force for the AC corrosion taking place on the steel surface at coating defects.

Stray current corrosion is caused by DC current, while AC corrosion is caused by AC current. All buried or immersed metallic structures are subject to DC stray current corrosion, while buried or immersed long pipelines laid in parallel with power lines are susceptible to AC corrosion.

AC corrosion mitigation is the process of designing and applying a pipeline grounding system to:

a. Prevent voltage spikes during fault conditions
b. Reduce AC current density to protect against AC-induced corrosion
c. Maintain AC step and touch potentials below 15 Vac to protect personnel from shock hazards

AC interference has several potential impacts on the safety of personnel and pipeline integrity. Isolation joints should be located at aboveground facilities such as terminals, pimping stations, offtakes, and instrument lines. The isolation joint can also be used to control the stray current by electrically separating sections of the pipeline. Isolation joints:

• Prevent the flow of electrical currents
• Prevent detrimental electrochemical interaction and improve the effectiveness of the cathodic protection system
• Electrically isolate the pipe system

Buried pipelines that run parallel to or in close proximity to power lines are subjected to induced voltages caused by the time-varying magnetic fields produced by the power line current. The induced electro-motive force causes current circulation in the pipeline and voltages between the pipeline and the surrounding earth. The currents flow through the pipeline because of the voltage difference induced on the pipeline and flow out through the damaged areas of the pipeline coating. This could cause AC corrosion in the pipeline, which could, in turn, lead to disastrous accidents such as a gas explosion or oil leakage. Therefore, pipelines should not be constructed (buried or aboveground) parallel to overhead power lines/overhead high-voltage power systems.

3.4.1 AC CORROSION MITIGATION APPROACHES

With reference to NACE SP0177-2014 the following AC mitigation strategies are suggested:

1. Fault shielding

    When pipelines are buried parallel to overhead power lines, in the event of a fault at the transmission power line tower, there could be a rapid discharge of fault current near the pipeline. This can result in coating damage. Fault shielding is a suitably designed grounding system that is installed between the tower footing and pipeline, and it acts to shield the pipeline and shunt harmful currents away from the pipeline by providing low-resistance path to earth. It is the form of a parallel shielding wire, either copper or zinc, connected to the pipeline.

2. Gradient control mats

    Install gradient control mat, and connect to the structure. Gradient control mat is a system of buried bare conductors, typically galvanized steel, copper, or zinc connected to the structure. It provides localized touch and step voltage protection by creating an equipotential area around the appurtenance.

3. Lumped grounding system

    Lumped pipeline grounding system consists of shallow or deep localized grounding conductors that are connected to the structure at strategic locations to reduce AC voltage levels along the pipeline. This provides protection to the structure during steady-state or fault conditions from the nearby electric transmission.

4. Gradient control wire

    It is the most common form of AC mitigation for high levels of AC-induced voltage along a pipeline. It functions the same as the lumped grounding system. Long continuous grounding conductors are installed horizontally and parallel to the pipeline. They are strategically located and sized to reduce the AC-induced voltage along the pipeline during steady-state or fault conditions from the nearby electric transmission.

3.4.1.1 Decoupler

Solid-state decouplers are used in conjunction with AC mitigation systems and are usually installed whenever the grounding system is connected to the pipeline.

A decoupler simultaneously performs two electrical functions: DC isolation and AC grounding. Under normal circumstances, the decoupler functions to isolate or block the flow of DC current (cathodic protection) to other equipment or grounding material, thereby eliminating any negative influence on your CP system caused by neighboring equipment or grounding materials. Yet the decoupler also provides safety grounding protection from AC faults and lightning.

AC fault protection: AC fault current generated by a failure of nearby equipment is a safety hazard. This AC current can be transmitted via the pipeline, endangering nearby workers or causing damage to your pipeline. When an AC fault occurs, the decoupler instantly conducts fault current to ground, carrying AC current away from the pipeline and providing instant protection.

Lightning protection: When a lightning strike occurs, it can present a safety hazard to personnel and pose a risk to the pipeline. The decoupler behaves in the same way as it does during an AC fault. It safely channels the high-voltage DC current of the lightning to ground, protecting nearby workers. As soon as this event has ended, the decoupler automatically switches back to its DC isolation/cathodic protection role.

Mitigation of induced AC current: Many pipelines are buried along a shared right-of-way with high-voltage power lines, which induce AC current in the pipeline. This is not only a safety hazard to workers, but can also cause corrosion issues. Decouplers such as the PCR or SSD are designed to continuously conduct induced AC current to ground, mitigating this risk while simultaneously isolating your CP system (Figure 3.38).

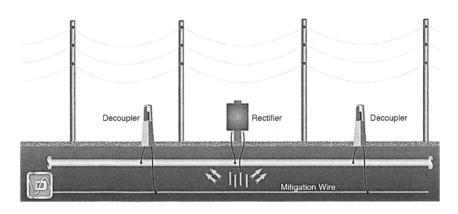

FIGURE 3.38 Visualizing how decouplers work (Dairyland 2020).

3.5 STRAY CURRENT CORROSION MITIGATION

Stray currents are constituents flowing in the electrolyte from external sources. Any metallic structure, for example, a pipeline, buried in the soil represents a low-resistance current path and is therefore fundamentally vulnerable to the effects of stray currents (Figure 3.39).

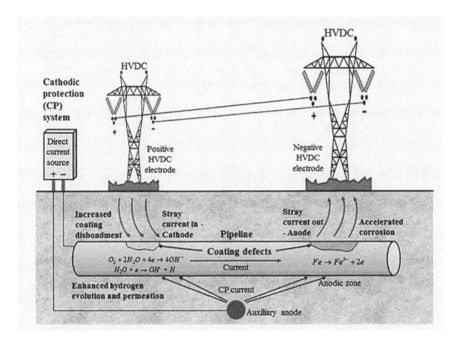

FIGURE 3.39 Stray current corrosion mechanism (Source: Frank Cheng, Shan Qian. 2017. *Accelerated corrosion of pipeline steel and reduced cathodic protection effectiveness under direct current interference*. Vol. 148, in *Construction and building materials*, by Shan Qian Frank Cheng, 675–685. ScienceDirect. Accessed January 8, 2020. https://advanceseng.com/a ccelerated-corrosion-pipeline-steel-reduced-cathodic-protection-effectiveness/).

3.5.1 STRAY CURRENT SOURCES

- HV DC and AC power transmission system
- Electrical train
- Impressed current cathodic protection (ICCP) system DC

3.5.2 STRAY CURRENT CORROSION PREVENTION

Stray current corrosion prevention is described in Sections 3.5.2.1–3.5.2.3.

3.5.2.1 Construction Technique

The pipeline shall not follow an overhead power line in parallel.

3.5.2.2 Corrosion and Prevention of DC Stray Current

a. The current out of design or regulated circuit is called stray current. If stray current flows into buried metal and then from metal into the earth or water, intensive corrosion will occur at the place where the current flows out. It is usually called galvanic corrosion. Its features are as follows:
   - Intensive corrosion.

- Corrosion is concentrated on the local.
- Corrosion is often concentrated on the defects of coating, if there exists corrosion-resistant coating.

Metallic pipelines that are interfered by stray current could have pitting corrosion in short time (Figures 3.40 and 3.41).

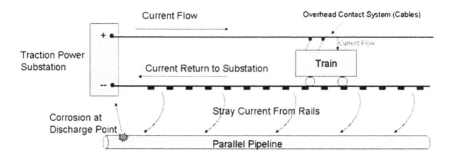

FIGURE 3.40   DC interference (Tong 2015).

FIGURE 3.41   Stray currents – AC interference (Tong 2015).

b. The most practical and effective way to prevent stray current interference is electric drainage method. Artificially lead interference current back to interference source or flow back to rectifier through regression line. This method requires connection between pipeline and regression line. This way of preventing pipeline galvanic corrosion is called electric drainage method. Common drainage methods include direct method, polarity method, compulsive method, and grounding method, as shown in Table 3.26.

TABLE 3.26
Drainage Protection Methods (Tong 2015)

Methods	Direct drainage	Polar drainage	Compulsive drainage	Grounding drainage
Schematic diagram				
Application conditions	1. There is a stationary anodic section on the interfered pipeline 2. Around the grounding electrode of the DC power supply station or negative regression line	Pipe-to-soil potential on the interfered pipeline has a positive and negative alternation	Small potential difference between pipeline and track.	Cannot directly conduct electric drainage to interference source

(Continued)

TABLE 3.26 (CONTINUED)
Drainage Protection Methods (Tong 2015)

Methods	Direct drainage	Polar drainage	Compulsive drainage	Grounding drainage
Advantages	1. Simple and economic 2. Favorable effect	1. Simple installation 2. Wide application range 3. Power-free	1. Wide protection range 2. Applicable to special occasions where other electric drainage methods can't be applied 3. Provide a cathodic protection when train car is out of service	1. Wide application range and is applicable to any occasions 2. Little interference to other facilities 3. It can provide part of the cathodic protection current (when using sacrificial anode grounding)
Disadvantages	Limited application range	Protection performance is poor when the pipeline is far away from the track	1. Accelerate galvanic corrosion of the track 2. There is great influence on the potential distribution of the track 3. Power supply is necessary	1. Less efficient 2. Auxiliary earth bed is necessary

3.5.2.3 AC Interference Hazards and Protection

3.5.2.3.1 Electric Field Effect

Pipeline–earth potential is slightly increased by the electrostatic field of the high-voltage lines and metallic pipelines, through coupling of distributed capacitance (see Figure 3.42). However, this effect is so small that it could be ignored.

3.5.2.3.2 Earth Electric Effect

If the pipeline is buried in the soil whose earth potential gradient changes greatly, the increase of the pipeline–earth potential caused is called earth electric field effect, mainly referring to the coupling phenomenon caused by current in soil. Grounding loop could be ignored during the normal operation of the high-voltage lines. However, when there is a fault, strong short-circuit current flows into the earth, increasing pipeline–earth potential and breaking down anticorrosive coating. Diverting potential generated after breakdown could damage people and equipment.

3.5.2.3.3 Electromagnetic Effect

Electromagnetic effect refers to the physical phenomenon caused by alternating electromagnetic field radiated by current carrying conductor cutting metal pipeline. Induced voltage and current in pipeline is the function of current, frequency, operating methods, and other factors of the high-voltage lines.

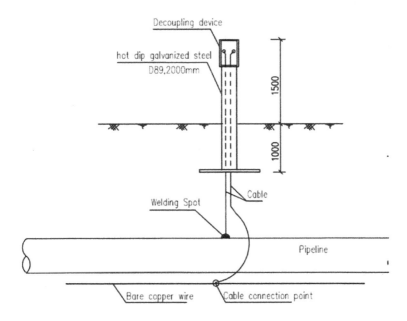

FIGURE 3.42 Stray currents – interference draining (Tong 2015).

3.5.2.3.4   Protection

The protection measures for buried pipelines are:

- Strengthen the anticorrosive coating quality of places where exists the earth's electric field interference
- Grounding treatment for ground pipelines or pipelines under construction
- Discharge induced AC on pipelines to the earth
- Conduct segmented insulation using insulating flange at places where induced AC is not easy to be discharged

3.6   STRESS CORROSION CRACKING

Stress corrosion cracking is a form of environmental-assisted cracking (EAC). The term "EAC" is used to describe all types of cracking in the pipeline that is influenced by the environment and stress. In a pipeline environment when water comes in contact with steel, there is potential for the minerals and gases to create an initial corrosion site that is acted on by stresses that result in crack growth. Contributing factors to crack growth are residual stresses, temperature, load stress, bending, and local stresses. If there is no stress, then crack growth will not occur and the result will be general wall thinning or pitting through corrosion (Ginzel and Kanters 2002).

The conditions that are present during crack growth dictate the type of SCC that may occur. For example, electrolyte pH will determine if the cracking is Intergranular or transgranular. High pH conditions cause intergranular cracks, while low pH conditions create non-classical SCC that results in transgranular cracks with mixed modes at the crack tip. The crack faces in low pH cracks also show evidence of secondary corrosion and appear wider than high pH SCC.

A typical example of SCC is:

- A 32-inch diameter gas transmission pipeline, north of Natchitoches, Louisiana, belonging to the Tennessee Gas Pipeline exploded and burned from SCC on March 4, 1965, killing 17 people. At least 9 others were injured, and 7 homes 450 feet from the rupture were destroyed.

Stress cracking of different alloys does occur depending on the type of corrosive environment. Stainless steels crack in chloride atmosphere. It has been found that cracks often initiate in trenches or pits on the surface, which can act as stress raisers. In some cases, crack initiation has been associated with the formation of a brittle film at the surface.

Major variables influencing SCC include solution composition, metal/alloy composition and structure, stress, and temperature (Natarajan 2012). The causes of stress corrosion cracking are:

a. A compelling cracking environment: the four factors controlling the formation of the potent environment for the initiation of SCC are the type and condition of the coating, soil, temperature, and cathodic current levels. The most common media where stress corrosion cracking occurs are

chloride-containing solutions, but in other environments, such as caustics and polythionic acid, problems with SCC is known to appear. Some environments that may cause SCC of steels of various grades are listed below. Environments where steels are prone to stress corrosion cracking:
- Seawater
- Condensing steam from chloride waters
- Polythionic acid (sensitized material)
- $NaOH–H_2S$
- Acid chloride solutions
- $NaCl–H_2O_2$
- H_2S + chlorides

b. Pipeline coating types: SCC normally starts on the pipeline surface at areas where coating disbondment or coating damage occurs. The resistance of a coating to disbonding is a primary performance property that affects all forms of external pipeline corrosion.

c. Soil: The soil type, drainage, carbon dioxide (CO_2), temperature, moisture content of the soil, and electrical conductivity form a conducive environment for the formation of stress corrosion cracks.

d. Cathodic protection: CP current forms a carbonate/bicarbonate environment at the pipeline surface, where high pH SCC occurs. For near-neutral pH, SCC CP is absent.

e. Temperature: temperature has a significant effect on the occurrence of high pH SCC, while it has no effect on near-neutral pH SCC. Elevated temperatures (normally above 60°C (140°F) for chloride-induced SCC, and about 80°C (176°F) in the presence of H_2S. Though it has been noted that cases exist where chloride-induced SCC has occurred at temperatures lower than 60°C (140°F).

f. Pipe material characteristics: a number of pipe characteristics and qualities such as the pipe manufacturing process, type of steel, grade of steel, cleanliness of the steel (the presence or absence of impurities or inclusions), steel composition, plastic deformation characteristics of the steel (cyclic-softening characteristics), steel temperature, and pipe surface condition are necessary conditions for the formation stress corrosion cracks.

g. Stress – when the tensile stress is greater than the threshold stress: if the tensile stress is higher than the threshold stress, this can lead to SCC, especially when there is some dynamic or cyclic component to the stress.

h. Pipe pressure: stress corrosion cracking is linked to the pressures exerted on the pipe. As the pressures within the pipe are increased, the growth rates for cracks also increase. The circumferential stress (hoop stress) generated by the pipeline-operating pressure is usually the highest stress component that exists.

3.6.1 MITIGATION OF STRESS CORROSION CRACKING OF PIPELINES

3.6.1.1 Material Selection
The first line of defense in controlling stress corrosion cracking is to be aware of the possibility of SCC at the design and construction stages. By choosing a material that

is not susceptible to SCC in the service environment, and by processing and fabricating it correctly, subsequent SCC problems can be avoided. Unfortunately, it is not always quite that simple. Some environments, such as high-temperature water, are very aggressive and will cause SCC of most materials. Mechanical requirements, such as a high yield strength, can be very difficult to reconcile with SCC resistance (especially where hydrogen embrittlement is involved).

The material selection process should reflect the overall philosophy regarding design life, cost profile, inspection and maintenance philosophy, safety and environmental profile, failure risk evaluations, and other specific project requirements (Singh 2014).

The material selected for sour service environment should be resistant to SCC and meet the requirements of the following standards:

- CSA Z662, Clause 16, Sour service
- NACE MR0175/ISO 15156 for upstream oil and gas process piping within the scope of ASME B31.3 – materials for use in H_2S-containing environments in oil and gas production
- NACE MR0103 for downstream refining and upgrading process piping within the scope of ASME B31.3
- NACE TM 0284-96, Evaluation of Pipeline and Pressure Vessel Steels for Resistance to Hydrogen Induced Cracking

NACE MR0175/ISO 15156 is an international standard that provides the requirements for metallic materials exposed to H_2S in oil and gas production environments. NACE MR0175 defines sour service as "Exposure to oilfield environments that contain sufficient H_2S to cause cracking of materials by the mechanisms addressed by NACE MR0175/ISO 15156".

Hardness and chemical composition control are the basis to mitigate the H_2S corrosion that could be formed in different forms of SSC (sulfide stress cracking corrosion), SCC, HIC (hydrogen-induced cracking corrosion), HISC (hydrogen-induced stress cracking corrosion), etc.

NACE MR0175/ISO 15156 standard addresses all mechanisms of cracking:

- Sulfide stress cracking (H_2S)
- Stress corrosion cracking (Cl–)
- Environmental cracking (synergistic, H_2S, and Cl–)
- Various hydrogen-induced cracking mechanisms

3.6.1.2 Environment

The most direct way of controlling SCC through control of the environment is to remove or replace the component of the environment that is responsible for the problem, though this is not usually feasible. Where the species responsible for cracking are required components of the environment, environmental control options consist of adding inhibitors, modifying the electrode potential of the metal, or isolating the metal from the environment with coatings.

The corrosivity of H_2S is a major risk for carbon steel, austenitic stainless steel and other CRA materials. Therefore, gas sweetening process unit should be designed and considered for gas processing plants with sour service.

3.6.1.2.1 Gas Conditioning Systems

Natural gas, whether produced from a condensate field or as associated gas from an oil reservoir, usually contains water vapor (H_2O) and frequently contains H_2S and/or CO_2 and heavy hydrocarbons. Other contaminants in the gas include CS_2, COS, mercaptans (RSH), N_2, O_2, Hg, solids-hydrates, asphaltenes, and dust.

Some or all of these contaminants need to be removed to meet gas specifications. This is done in the gas conditioning module. The gas conditioning module is usually installed at the inlet of the gas processing plant.

3.6.1.2.1.1 Inlet Separation Inlet separation removes any remaining water and heavy hydrocarbons from the gas stream.

3.6.1.2.1.2 Gas Treating (Sweetening) Gas treating is used to reduce the "acid gases", carbon dioxide (CO_2) and hydrogen sulfide (H_2S), along with other sulfur-containing compounds, to sufficiently low levels to meet contractual specifications or to permit additional processing in the plant without corrosion and plugging problems.

One of the most dangerous compounds present in natural gas is H_2S. Its removal is necessary to reduce corrosion. H_2S gas is highly toxic. In addition, CO_2 creates a lot of technical and safety problems if the concentration is higher than the imposed limit according to sales gas quality requirements. CO_2 forms a strong acid that is highly corrosive in the presence of water. Carbon dioxide is also non-flammable, and, consequently, large quantities reduce the heating value of the fuel.

Removal of H_2S and CO_2 from natural gas is called "gas and liquid sweetening". Many such sweetening technologies have been developed. H_2S and CO_2 can be removed from gas streams by physical absorption, chemical reaction and adsorption, or combinations of these processes. A common way to remove H_2S and CO_2 from natural gas is by using a chemical solvent–amine system, which uses a contact tower with trays or structured packing to pass the sour gas through the amine liquid, thus absorbing the H_2S and some of the CO_2. See the principle of this process in Figure 3.43.

3.6.1.2.1.3 Other Conditioning Process Selective removal of free gases like N_2, He, Ar, and Ne from natural gas is done using a special type of molecular sieve (Figure 3.44).

When aluminum heat exchangers and equipment are used, mercury removal from gas is often necessary. It is typically done by passing the gas through a bed of sulfur-impregnated activated charcoal or alumina where the mercury reacts to form mercuric sulfide, H_2S.

Filters are used to remove solid particles.

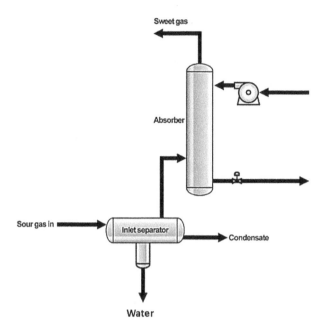

FIGURE 3.43   Typical amine gas treating.

3.6.1.3 Stress

Stresses in the pipe may lead to premature degradation of the pipeline strength. Stresses acting on the pipe include:

* Residual stress from the manufacturing process
* External stress such as those incurred due to bending, welding, mechanical gouges, and corrosion
* Secondary stresses due to soil settlement or movement

One of the causes of stress corrosion cracking is the presence of stress in the pipe components; therefore to mitigate SCC, you have to eliminate stress or reduce it below the threshold stress for SCC. If SCC is being caused during welding or forming, then use stress-relief annealing to relieve stress.

Partial stress relief around welds and other critical areas may be of value to large structures for which full stress-relief annealing is difficult or impossible. Stresses can also be relieved mechanically. For example, hydrostatic testing beyond yield will tend to "even-out" the stresses and thereby reduce the peak residual stress.

Laser peening, shot-peening, or grit-blasting can be used to introduce a surface compressive stress that is beneficial for the control of SCC. These processes can be performed uniformly.

3.6.1.4 Coating

Clean and prepare the pipe surfaces through shot-peening (a cold work process used to finish metal parts to prevent fatigue and stress corrosion failures and prolong product life for the part it entails striking a surface with shot (round metallic, glass, or ceramic particles) with the force sufficient to create plastic deformation) and then apply special coatings such as fusion-bonded epoxy to protect the pipeline from the occurrence of SCC.

3.6.1.5   Cathodic Protection

Use cathodic protection systems to protect pipe from corrosion. Perform annual surveys to check the effectiveness of cathodic protection. Repair areas with pipe-to-soil potential below −0.85 V.

Methods of preventing corrosion and SCC on existing pipelines include minimizing the operating temperature and controlling the CP levels to values more negative than −850 mV CSE. Minimizing pressure fluctuations on operating pipelines is also effective in preventing SCC initiation. A more detailed discussion of coatings and cathodic protection is given in Section 3.2.

3.7   MITIGATION OF HYDROGEN-INDUCED CRACKING

Failures due to environmental cracking mechanisms related to hydrogen such as HIC, hydrogen stress cracking (HSC), stepwise cracking (SWC), and stress-oriented hydrogen-induced cracking (SOHIC) are not addressed in detail in this chapter.

Specification and use of materials manufactured with demonstrated HIC resistance is the preferred method for preventing failures by this mechanism. However, many of the preventative measures described in this chapter can help mitigate failures by HIC.

For more information on requirements to prevent HIC failures refer to:

- NACE MR 0175/ISO 15156, petroleum and natural gas industries – materials for use in H_2S-containing environments in oil and gas production
- NACE TM 0284-96, Evaluation of Pipeline and Pressure Vessel Steels for Resistance to Hydrogen Induced Cracking
- CSA Z662, Clause 16 Sour Service

3.8   MITIGATION OF SULFIDE STRESS CRACKING

Failures due to environmental cracking mechanisms such as sulfide stress cracking are not specifically addressed in this chapter. Selection of materials resistant to SSC and control of combined stress are considered the primary acceptable means to prevent failures by this mechanism. Stress can result from welding, installation, soil loading, thermal expansion, operating pressure, defect, etc. (CAPP 2018).

For more information on requirements to prevent SSC failures refer to:

- NACE MR 0175/ISO 15156, petroleum and natural gas industries – materials for use in H_2S containing environments in oil and gas production
- CSA Z662, Clause 16 Sour Service

3.9 HOW TO FIND CORROSION

Internal corrosion of pipelines is not normally a problem as a treatment to remove the corrosive elements within the product is usually undertaken at the inlet station/process plant. Where corrosive elements are known to exist and treatment is only limited or not possible, e.g. Wellsite lowliness, unestablished crude oil, etc., steels manufactured to NACE MR0175-91 and other appropriate standards may be used.

Notwithstanding the above as an additional safeguard, internal corrosion monitoring of the pipeline may be undertaken. This monitoring could take the form of corrosion coupons and/or electrical resistance probes installed and monitored at each end of the pipeline.

Additionally, facilities may often be provided for the running of intelligent pigs.

External corrosion may be monitored by assessing the performance of the impressed current cathodic protection system.

To find corrosion in oil and gas pipelines, the following four inspections are proposed.

    a. Visual inspection
    b. Pipe-to-soil readings
    c. In-line inspections
    d. Drone/robot technology

The external protection of the pipeline and ancillary facilities is achieved by a system comprised of a coating and cathodic protection. The performance of this system is assessed by regular monitoring of pipe-to-soil potentials at selected intervals along the pipeline. From these results, conclusions may be drawn concerning the level of cathodic protection being achieved and by the performance of the coating.

3.9.1 INTERNAL INSPECTION

Pipelines shall be internally inspected utilizing an in-line inspection tool. A series of pigging runs, including cleaning and geometric and profile pigs, shall be used to ensure the safe passage of the inspection pig. A magnetic flux leakage, or ultrasonic pig, may be used to detect metal loss in pipe walls and will locate and size areas of corrosion and mechanical damage. The magnetic flux leakage pig can also detect circumferential cracks.

For pipelines that may be subject to stress corrosion cracking, consideration should be given to the use of an elastic wave or transverse flux leakage internal inspection tool.

3.9.1.1 In-line Inspection

In-line inspection refers to a preventative maintenance examination of pipelines to identify corrosion, cracks, and other defects that may result in catastrophic failure of the structure. It is a form of non-destructive examination.

There are many pipelines in the world built 50–60 years ago, still operating effectively, although many of these have been repaired and rehabilitated over their lifetime. With modern inspection techniques available, i.e. geo pigs measuring pipeline orientation and intelligent pigs measuring wall thickness and identifying cracks, the operating life of a pipeline can be extended well beyond the original design life until the cost of maintenance or repair becomes economically unviable.

In-line inspection tools are used to identify, locate, and measure anomalies. An instrumented vehicle/tool called "pig" is propelled through the pipeline and provides highly detailed information about the interior and exterior of the pipeline. These tools can identify changes in the wall thickness of the pipeline. Any changes in wall thickness identified can be classified based on the calibration of the inspection tool on pipe sections with known features. The accuracy of these inspection tools is steadily improving. However, there are limitations in the measurement of feature size (length and depth), feature circumferential position within the pipeline, and feature location along the pipeline (Roland and Dominic 2000).

There are three types of anomalies looked for:

- Metal loss (or corrosion)
- Cracks
- Changes to the shape of the pipe

The pipeline must allow for the passage of the tool and have specific equipment in place to handle the equipment required.

Based on the test results each anomaly must be assessed to determine if it could cause damage. The company would then address significant issues and make required repairs to the pipeline. Many types of defects can be assessed using the correct in-line inspection tools. Magnetic flux leakage tool is an example of an inspection tool. It uses magnetic technology to detect, discriminate, and size metal loss in the liquid and gas pipeline.

3.9.1.1.1   Standards
- NACE recommended practice: RP102-2002, in-line inspection of pipelines
- API 1163, in-line inspection systems qualification standards
- ANSI/ASNT ILI-PQ-2005, in-line inspection personnel qualification and certification (qualifies vendor personnel including analysts)
- 49 CFR 192 – subpart O (for gas transmission)
- 49 CFR 195 (for liquid transmission)
- Your company standards

3.9.1.1.2   Smart Pig
Intelligent pigging is an inspection method whereby an inspection probe (also known as a smart pig) is propelled through a pipeline. When the smart pig travels along the pipeline, it gathers information or data such as pipeline's diameter, curvature, bends, temperature, anomalies on the pipe internal wall, metal loss, general corrosion,

erosion corrosion, pitting, weld anomalies, hydrogen-induced cracking, and the presence and location of corrosion.

The advantage of the smart pig over the traditional method of pipeline inspection is that:

1. Smart pig is able to clean and inspect the pipeline while it travels through the pipeline.
2. It allows pipelines to be cleaned and inspected without having to stop the flow of the product.
3. Smart pig saves companies both time and money because it provides cleaning and inspection services at the same time.
4. Some smart pigs are equipped with GPS that can help in mapping a pipeline. This helps maintenance personnel to save time and money by showing the exact location of a defect, instead of having to excavate a large area to reach a specific location in the line.

FIGURE 3.44   A smart pig.

Smart pigging is a non-destructive evaluation technique which uses the following methods of in-line inspection technology:

a. Ultrasonic testing (UT)
b. Magnetic flux leakage (MFL)

3.9.1.1.3   Pigging Facilities

In order to carry out internal pipeline cleaning and inspection, conventional pipeline pigging facilities should be provided at the respective ends of each pipeline (see Figures 3.45 and 3.46).

The barrels of the launcher and receiver should be of the in-line type with safety interlock end closure doors, each barrel being of sufficient length to accommodate internal cleaning devices (PIGS). The pig traps should be designed in accordance

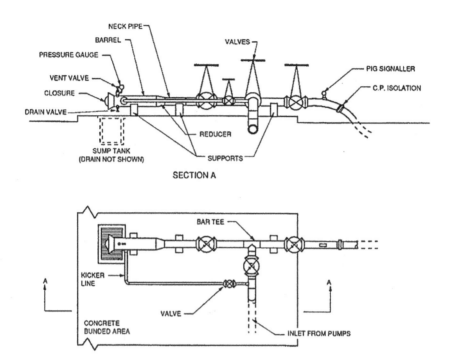

FIGURE 3.45 Typical pig launching facilities (Source: HRR. 1992. *Onshore pipeline design base manual.* Design guideline, JP Kenny, unpublished).

with the relevant codes of practice for pressure vessels (e.g. BS 5500). Sphere/barred tees should be fitted into the pipeline to allow the free passage of the pigs at all locations where the branch-to-line ratio exceeds 40%.

Double valve isolation should be provided on flow, isolation, and kicker/bypass lines; normally these will be ball valves on piggable sections and gate valves on non-piggable sections. The method of operation for these valves will depend on their size and the operation philosophy for the pipeline having due consideration to plant emergency shutdown (ESD) requirements.

Pig launching and receiving is normally a manual field operation, and detectors (pig signallers) are usually installed to indicate the free passage of the pigs within the system.

With reference to the pig launcher and receiver, the hinged end closure should be provided with positive pressure interlocks to prevent the end being opened while the barrel is under any pressure whatsoever. The mainline bypass and kicker line valves are usually all interlocked with visual instrumentation, e.g. pressure gauges, being provided, allowing only pre-determined valve operation sequences.

A full range of secondary branches of suitable size should be provided on the launcher/receiver to accommodate standard facilities, i.e. vents, drains, purge, safety instrumentation, and washdown requirements. No branch connection should be less than 1-inch nominal bore and should in all cases be provided with at least one isolating valve.

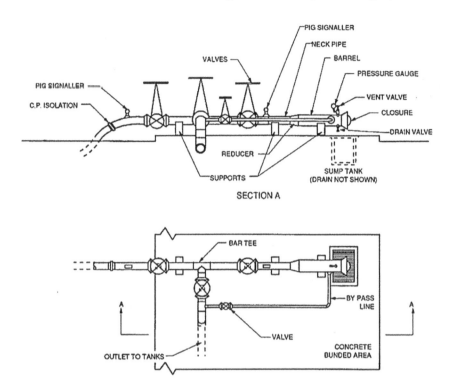

FIGURE 3.46 Typical pig receiving facilities (Source: HRR. 1992. *Onshore pipeline design base manual.* Design guideline, JP Kenny, unpublished).

Bend radii associated with pig launcher/receivers should be compatible to the use to which the line is to be put with particular regard to the use of intelligent pigs and the minimum radius imposed thereby. Launch and receipt facilities should be sited on a paved area, which includes either a sump of adequate size to retain any liquid spillage when the trap is opened or kerbed to provide a bounded area sufficient to retain any liquid spillage.

Consideration should be given to the need for platforms and walkways necessary for the operation of the facility, particularly on larger-diameter pipelines.

3.9.1.1.4 In-line Inspection Technologies
Methods of in-line inspection technology:

 a. Ultrasonic testing
 b. Magnetic flux leakage

3.9.1.1.4.1 Magnetic Flux Leakage
- Most common method in use
- Longitudinal saturation
- Changes in flux field are interpreted

- Metal loss (corrosion) is reported as the percentage of wall loss
- Other pipe abnormalities are also reported
- It can run in dry or wet pipes
- May have to run chemical cleaning of the pipe
- Speed sensitive
- Permanent magnet's ability limits wall thickness to about 0.75 inch (19.28 mm)
- Standard resolution (±15%, 20% confidence factor)
- High resolution (±10%, 20% confidence factor)
- Magnetic flux leakage tool specificationsBased on the direction of magnetization, at least two types of tools are available
  - The standardized MFL tool that magnetizes the pipe wall in the axial direction has limited sensitivity to axially aligned defects.
  - MFL tools that magnetize the pipe wall in the circumferential direction are more sensitive for axially aligned metal loss but are likely to have different specifications.
- Direction of magnetization: axial/circumferential/diagonal
- The magnetic field strength H in Am-1 to meet the given POD and accuracy
- Axial sampling frequency or distance
- Nominal circumferential spacing of ID/OD discriminating sensors (if present)
- Location accuracy of the features with respect to the upstream girth weld, the upstream marker, and the orientation in the pipe
- If crack detection is possible and is included in the inspection scope of work, the contractor shall provide the following parameters:
  - Minimum depth, length, and the opening dimension of a crack to be detectible
  - The confidence level for the detection of this minimum crack
  - The accuracy of sizing of crack length and depth
  - The confidence level for sizing performance

The complete technical details/datasheet of the MFL tool must be given by the bidder.

3.9.1.1.4.2 *Ultrasonic Thickness*

- An array of UT heads transducers covers the surface of the pipe
- Liquid media is required as a couplant
- Speed sensitive
- No limits on wall thickness
- More expensive than MFL

Ultrasonic Thickness Tools Specification

UT-metal loss detection tool specification shall include:

- Axial sampling frequency or distance
- Nominal circumferential spacing of measuring sensors

- Diameter/dimensions of UT transducers
- Frequency of UT transducers
- Stand-off distance of UT transducers

UT-crack detection tool specification shall include:

- Tool length, weight, and number of bodies
- Axial sampling frequency or distance
- Nominal circumferential spacing of measuring sensors
- Dimensions of UT transducers
- Frequency of UT transducers
- The angle of UT signal in steel
- The direction of the angle of UT signal relative to pipe axis (i.e. longitudinal direction is 0°; circumferential is 90°)
- Minimum depth and length for a crack to be detectable
- The confidence level for the detection of this minimum crack
  - The accuracy of sizing of crack length and depth
  - The confidence level for the sizing performance

Benefit of UT
- Only requires direct access to one side of the test piece
- Can penetrate thick materials
- Accurately measures wall thickness and flaw depth
- Determines the maximum allowable operating pressure and estimates remaining service life based on ASME B31G, API 653, API 510, and other similar codes

3.9.2 EXTERNAL INSPECTION

If the use of an internal in-line inspection tool is not feasible, then an external aboveground survey should be undertaken. The following external survey methods may be used.

1. Close-Interval Potential Survey

    The CIPS survey determines the actual level of cathodic protection being experienced along the pipeline by measuring the pipe-to-soil potential and hence corrosion protection levels and quality of coating.
2. Direct Current Voltage Gradient Survey

    The DCVG survey shall be used for a detailed assessment of the condition of the pipeline coating. DCVG shall be used for coating defect size evaluation, defect length evaluation, defect corrosion status, and defect influence regarding electrical interference.
3. Pearson Survey

    The Pearson survey shall also be used to locate coating defects. It is very effective in tracing discontinuities or damage in buried pipelines' coating, as well as loose electrical contacts and their exact site on the pipe, allowing the prevention of major failures.

4. Visual inspection
5. Drone/robot technology

3.9.2.1 Visual Inspection

A visual inspection of all aboveground metallic facilities (e.g. pipe bodies, pipe supports, pipe fittings, aboveground pipe transitions, valve, and valve bodies). Signs of corrosion include discolored and/or peeling paint, evidence of rust or oxidation of the metal surface, and any other condition that may require remedial action.

The experienced inspector can often determine the type of corrosion, such as general corrosion, pitting corrosion, crevice corrosion, weld- and heat-affected zone corrosion, and erosion corrosion from visual inspection. The degree of corrosion can be measured and described and documented by the use of sketches or photographs. For exact measurements of local corrosion penetration caused, for example, by pitting corrosion, various types of mechanical or optical measuring instruments can be used. Material thinning due to general corrosion may be difficult to determine exactly, without the use of additional non-destructive inspection methods, such as ultrasonic. Initial cracks caused by stress corrosion or corrosion fatigue are often difficult to detect visually. If such defects are likely, then methods to make the cracks visible are needed (magnetic particles, liquid penetrant, eddy current). Mirrors, boroscopes, flexible fiber-optic instruments, or small video cameras can be used together with light sources, to look inside small pipes or narrow spaces in the equipment.

3.9.2.1.1 Direct Assessment of the Pipe

Direct assessment: a structured process for pipeline operators to assess the integrity of buried pipelines. Inspection results are validated through integrity digs where you excavate the pipeline to evaluate it.

- Direct assessment is one of the three acceptable methods for evaluating the integrity of a pipeline segment.
- Direct assessment may be used either as a primary or supplementary method, implemented in conjunction with one of the other primary assessment methods, i.e. in-line inspection or hydrostatic pressure testing.
- Direct assessment (DA) is limited to evaluating the risks of three time-dependent threats to the integrity of a pipeline segment:
  - External corrosion
  - Internal corrosion
  - Stress corrosion cracking
- When a pipeline segment is scheduled for a full integrity re-assessment at an interval longer than 7 years, confirmatory direct assessment (CDA) may be used during the seventh year following a baseline assessment to verify or "confirm" the integrity of a pipeline from external and internal corrosion threats only.
- If ECDA finds pipeline coating damage, the operator must integrate the data from ECDA with one-call notification information and right-of-way information to evaluate the segment for the threat of third-party damage.

Direct assessment may be used to evaluate the integrity of pipelines for the following reasons.

- Where ILI or hydrostatic pressure testing cannot be used
- To avoid impractical, costly retrofitting of a pipeline
- To avoid interrupting gas supply to a station fed by a single pipeline
- To provide an alternative where sources of water for hydrostatic pressure testing are scarce, and where water disposal may create problems
- DA may provide a more effective, equivalent alternative to ILI and hydrostatic pressure testing for evaluating a pipeline's integrity.

3.9.2.1.1.1   How Is Direct Assessment Carried Out?   External Corrosion Direct AssessmentAccording to NACE standard RP0502-2002, an external corrosion direct assessment is a four-step process that combines pre-assessment, indirect inspection, direct examination, and post-assessment to evaluate the threat of external corrosion to the integrity of a pipeline. ECDA requires:

1. Step One: Pre-assessment – to gather and integrate data to determine the feasibility of using ECDA for a segment, the identification of ECDA regions, and the identification of two indirect examination tools to be used on the ECDA region.
2. Step Two: Indirect examination – to evaluate the pipe segment and identify indications of potential external corrosion, to classify the severity of those indications, and determine urgency for their excavation and direct examination.
3. Step Three: Direct examination – to examine the condition of the pipe and its environment, to determine actions to be taken should corrosion defects be found, and to identify and address root causes.
4. Step Four: Post-assessment – to determine a segment's remaining life, its re-assessment interval, and the effectiveness of using ECDA as an assessment method.

Internal Corrosion Direct Assessment
Internal corrosion direct assessment is a process an operator uses to identify any area along a pipeline where fluid or other electrolyte introduced during normal operation or by an upset condition may reside. The process identifies the potential for internal corrosion caused by microorganisms, fluid with carbon dioxide, oxygen, hydrogen sulfide, or other contaminants present in the gas.

For internal corrosion direct assessment (ICDA), there are four-step processes, based on the principle that liquids collect on the bottom of a pipe when a "critical angle of inclination" is exceeded for a specific gas flow velocity:

1. Step One: Pre-assessment – to gather and integrate data and information to determine whether ICDA is feasible for the segment, to support use of a model to identify locations where liquids may accumulate, and to identify where liquids may enter the pipeline.

2. Step Two: Corrosion Direct Assessment (CDA) region identification – to apply a specific model to identify elevation conditions and other pipeline fittings where liquids may accumulate.
3. Step Three: Direct examination – to excavate and examine pipe locations identified by the process as most likely for internal corrosion, and to evaluate the severity of defects and remediate as code requires.
4. Step Four: Post-assessment evaluation and monitoring – to evaluate the effectiveness of the ICDA process, to monitor segments where internal corrosion was identified, and to determine re-assessment intervals.

Stress Corrosion Cracking Direct Assessment
Stress corrosion cracking direct assessment (SCCDA) requires a plan that provides for:

1. Data gathering and integration: to determine whether the conditions for stress corrosion cracking are present, requiring an assessment for SCC; to prioritize pipeline segments for assessment; and to gather and evaluate data related to stress corrosion cracking at all operator excavation sites. When all of the following conditions for high pH SCC are present, an assessment method must be applied.

   • Operating stress greater than 60% of SMYS
   • Operating temperature greater than 100°F
   • Within 20 miles downstream of a compressor station
   • Age greater than 10 years
   • Pipe coating other than fusion-bonded epoxy

2. Assessment method: to evaluate segments for the presence of stress corrosion cracking; determine its severity and prevalence; repair, remove, or hydrostatically test the valve section; and determine any further mitigation requirements.

Should conditions for SCC be present in a segment, the segment must be assessed and remediated, as specified in Appendix A3 of ASME B31.8S, applying:

• The bell hole examination and evaluation method
• The hydrostatic pressure testing method for SCC

Applying CDA requires a plan specifying that CDA can only be used on internal and external corrosion threats.

1. For external corrosion (EC), the plan must comply with 49 CFR Part 192.925, however:
   • Only one indirect examination tool may be used, and one high-risk indication is examined in each ECDA region.
   • All immediate indications must be excavated in each ECDA region.

2. For internal corrosion (IC), an operator's plan must comply with §192.927, however: only one high-risk location must be excavated in each ICDA region.

3. When applying either ICDA or ECDA, if defects are found requiring remediation before the next scheduled assessment, the operator is required to apply the formula in §6.2 and 6.3 of the National Association of Corrosion Engineers Recommended Practice 0502 to schedule the next assessment.

3.9.2.1.1.2 Reference Standards In December 2004, NACE adopted Recommended Practice 0204 for Stress Corrosion Cracking Direct Assessment and is now in the process of adopting a proposed recommended practice for internal corrosion direct assessment. These will provide operators additional guidance for addressing threats.

- For ECDA: NACE RP 0502-2002, ASME B31.8S §6.4, and LI 2189
- For ICDA and for SCCDA: ASME B31.8S § 6.4, Appendices A2, B2, and A3
- For ICDA: 49 CFR § 192.927, NACE standard SP0206-2016

3.9.2.2 Pipe-to-Soil Readings

Cathodic protection systems, once installed, require maintenance and inspection in order to identify any areas suffering under protection and can identify areas where increased current demand indicates coating damage.

1. Pipe-to-soil readings shall be taken anytime that buried or submerged pipe coating is exposed:
   - Coating is removed or damaged
   - Not required when vacuum excavated
2. Adequate cathodic protection
   - The level of cathodic protection shall be considered adequate when the minimum pipe-to-soil potential is at least a −0.85 V (negative 850 mV) for metallic pipelines. This is the level at which metallic pipe no longer corrodes.
   - A 100-mV shift is accomplished after the CP has been turned off (may require 2 minutes to 24 hours).
3. Inadequate cathodic protection
   Reading less than − 0.85 V (more positive than) may indicate inadequate CP, requiring medial action to correlut the situation.
4. Excessive cathodic protection
   - The amount of cathodic protection must be controlled so as not to damage the protective coating or the pipe. This is accomplished by limiting the maximum "on" pipe-to-soil potential to negative (−) 2.5 V.
   - Any reading greater than −2.5 V indicates excessive CP requiring remedial action to correct the situation.

5. CP records shall be maintained for no less than 5 years. It is recommended that these CP records be maintained for the life of the pipeline.

Prior to taking any pipe-to-soil readings it is important to check or calibrate the reference electrodes being used. This can be undertaken, as shown in Figure 3.31. The test is simply to place the porous plugs of a standard (unused) electrode and the field electrodes end to end and measure the millivolt difference. Generally, if the difference is less than 4–6 mV, no maintenance of the electrodes will be required.

Testing of the field reference electrodes should be undertaken each morning prior to the start of the survey. The millivolt difference and polarity between the working electrodes and the standard should be recorded.

3.9.2.2.1   Measurement of Pipe-to-Soil Potential

Monitoring of cathodic protection potentials on onshore pipelines is carried out at the test post facilities installed along the pipeline route using a portable copper/copper sulfate reference cell and a high impedance voltmeter.

3.9.2.2.1.1   Close-Interval Potential Surveys   Close-interval potential surveys or close-interval surveys are used to determine the level of cathodic protection being experienced along the pipeline by measuring the pipe-to-soil potential and hence corrosion protection levels, as shown in Figure 3.47. There are three types of close-interval surveys: on/off, depolarized, and on surveys.

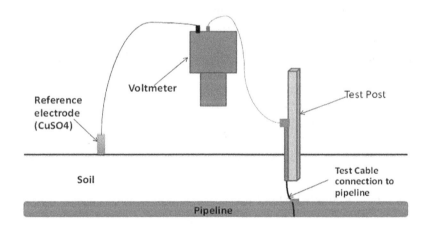

FIGURE 3.47   Close-interval potential surveys.

Procedure

a. Pipe-to-soil readings are taken utilizing a copper–copper sulfate half-cell and a voltmeter.

b. To ensure a proper pipe-to-soil potential reading, remove the cap and place the porous plug of the copper/copper sulfate reference electrode in firm contact with the earth over the pipeline or close to it.

c. This may require "digging in" where the earth's surface is dry. In dry areas, it is necessary to moisten the earth around the electrode with freshwater to obtain a good contact.

d. The red lead/wire from the digital voltage meter is connected to the structure/pipeline (via test point terminals or direct contact with pipeline); the black lead from the digital voltage meter is connected to the half-cell; and the pipe-to-soil potential is read and recorded.

e. Some half-cells utilize the voltmeter attached directly to the top of the half-cell and require only on lead to be attached to the pipeline facilities or test lead to be tested.

f. The voltmeter must be of high-input impedance to ensure accuracy.

3.9.2.2.1.2   Remote Monitoring   RMUs are available to remotely monitor the output of transformer–rectifier units and drain point potentials. Remote monitoring reduces the cost related to conveying engineers to remote sites. RMUs can be used for remote monitoring and wireless transmission of corrosion and cathodic protection data over long distances.

3.9.2.2   Criteria for Cathodic Protection

For onshore pipelines/structures, based on a field survey and laboratory test of the soil samples, a decision can be made on the required minimum pipe-to-soil protection criteria. For aerobic soils, the protection criteria would be −0.85 V with respect to $Cu/CuSO_4$ reference electrode. For anaerobic soils containing sulfate-reducing bacteria, the protection criteria would be −0.95 V with respect to $Cu/CuSO_4$ reference electrode.

A pipe-to-soil potential of −0.85 V versus a $Cu–CuSO_4$ electrode indicates satisfactory cathodic protection.

The −0.85 criteria, as well as a potential shift of −300 mV versus a $Cu–CuSO_4$ electrode, is used for bare pipelines. It is also suggested that a −100 mV shift of potential in bare pipelines indicates a good degree of cathodic protection.

For high-strength steels with tensile strength in the range of 700–800 MPa, typical protection criteria would be −1.0 V with respect to $Cu/CuSO_4$ reference electrode (Table 3.27).

3.9.2.3   Drone/Robot Technology

Due to the difficulty in accessing most piping within process refineries, chemical plants, and many outdoor pipelines, it can be impossible to inspect with conventional in-line inspection methods. During turnarounds, these assets need to undergo thorough inspections, as the hazardous materials that flow through them create prime environments for various damage mechanisms.

To prevent the need for confined space entry (CSE), you can conduct inspections with robotics equipped with high-resolution cameras, UT and infrared scanning capabilities, and eventually high-resolution laser scanning.

Aboveground storage tanks throughout their operating life are subjected to considerable operational and environmental forces and subject to corrosion and cracks on the surface and subsurface level that travels parallel to the surface. Usually cranes,

TABLE 3.27

Explanation to the Readings from Cathodic Protection Test

Voltmeter reading	Explanation (what the voltmeter reading indicates)
Greater than −0.88 V	These sections of the pipeline are adequately protected.
−0.85−−0.88 V	The pipe structure at those sections still meets the standard for corrosion protection, but there is not much of a safety cushion. We have to monitor those sections of the pipeline with such reading closely to determine the rate at which the voltage is dropping and plan on adding anodes or performing other work on the system in the not-too-distant future.
Less than −0.85 V	The pipe structure at these sections does not meet the −0.85 V standard for corrosion protection and is out of compliance with regulatory requirements (refer to Clause 5.3.2.1 of ISO 15589-1). Possible reason for failure to achieve −0.85 V is excessively dry soil around the test point. If the pipes backfill was damp but −0.85 V reading still could not be measured, then there is the need to research the installation procedures to see if we can discover any clues. Break in a continuity bond or increased resistance between the point of connection and the point of test due to a poor cable connection. Deterioration of or damage to the pipeline protective coating. Reversed connections at the transformer–rectifier, which is a very serious fault that could result in severe damage to the pipeline in a relative short period.
−0.4−−0.6 V	It means the steel pipe has no cathodic protection or that the anodes are completely shot. There is a need for investigation by a corrosion engineer.

scaffolds, rope-access, and people are used to inspect both the storage tanks and the overall structure. Maintenance and inspection are, thus, costly, time consuming, and risky for those who carry out the inspection. The solution to this can be to let autonomous drones do the work.

Inspection robots play an important part in the oil and gas industry by taking the invaluable role of inspection, monitoring, and surveillance of complex structures in the industries and, thereby, averting any disasters that may occur. The use of robots helps in reducing human intervention, increasing operational efficiency, reducing costs, and improving safety.

3.9.3 MANDATORY MONITORING

Mandatory inspections and monitoring that shall be conducted on a regular basis are:

1. Cathodic protection surveys – It is desirable to take a series of electrical measurements on a newly installed buried cathodic protection system to determine the initial level of cathodic protection. After the initial series, measurements should be made after six months and one year of operation.

This will enable the corrosion engineer to identify deficiencies and program corrective action. After the first year of operation, measurements should be made at least annually, unless conditions indicate more frequent testing (refer to Clause 11.6.4 of IGE/TD/1 edition 4).

The effectiveness of CP systems shall be determined by comparing survey results to criteria listed in Clause 6.2 of NACE SP0169-2013 or Clause 5.3 of ISO 15589-1. Any deficiencies or repairs which are outlined in the annual report and compromise the effectiveness of the CP system should be addressed within 12 months of discovery.

2. Rectifier monitoring/maintenance – all rectifiers and critical current sources shall be inspected at least once every two months for proper operation. Monthly inspections of cathodic protection rectifiers must be made to make sure that the units are operating as intended for their specific purpose. The inspections should include measurements of current and voltage output and a check of components for damage or deterioration.
3. Annual internal corrosion review – once per calendar year, the internal corrosion susceptibility of each licensed pipeline shall be reviewed by assessing one or all of the following data types:
   • Production and Operating Parameters
   • Gas Analysis
   • Corrosion Monitoring Data
   • Mitigation Program Records
   • Inspection and Repair Records

3.9.3.1 Supplemental Integrity Monitoring

1. Corrosion monitoring devices – pipelines which display proven internal corrosion activity, or are deemed as high-risk to internal corrosion activity, shall be monitored via corrosion coupons, probes, non-destructive testing, or other means of assessing pipe wall condition. Data from newly installed monitoring devices shall be collected at intervals not exceeding six months until a new collection frequency is determined by historical corrosion rates.

Adjustive surveys of cathodic protection systems are performed annually and include the following

• Complete inspection of rectifier and groundbed installation for the purpose of determining overall condition and efficiency with corrective measures taken as required
• Complete inspection of all resident pipe-to-soil meters for the purpose of determining condition and accuracy. Repair and recalibrate as required
• Rectifier output levels adjusted to pre-determined levels and pipe-to-soil potentials measured at provided pipeline test leads and resident pipe-to-soil meters. Faulty test leads discovered during the survey are repaired as

required. Rectifier output levels are re-adjusted during pipe-to-soil potential the survey to ensure pipe-to-soil potentials are maintained at adequate and balanced protective levels on all pipelines and in their entirety.

All cased road crossings are inspected, and measurements are taken to determine if contact between carrier pipe and casing is evident, and, where there is evidence of contact, corrective action is taken as required.

- Evidence of foreign interference is investigated and corrective measures implemented
- Inspection and repair or replacement of faulty insulating material or fittings as determined from pipe-to-soil potential measurements
- Completed and detailed records of all adjustive survey data (repairs and maintenance, measurements, settings, adjustments, etc.) are compiled and recorded for reference

When the pipe-to-soil potential survey indicates inadequate protection, check at the rectifier unit. The current output of the rectifier should be adjusted:

- If the current is high and accompanied by low voltage, suspect the pipeline and investigate for a possible short circuit to another metal structure.
- If the current output is low, with normal or high voltage, suspect the anode groundbed or connecting cables.

3.10   WHAT TO DO WHEN CORROSION IS FOUND

The integrity of the pipe is essential to the safe operation of the pipeline. To maintain the integrity of the pipeline, anomalies found on or within the pipe must be evaluated and, if necessary, repaired. The main objective of the repair process is to return the pipeline back to its original integrity or to ensure that the anomaly is left in such a state that the pipeline can safely maintain operation.

Repairs to the pipeline may be required because of several different circumstances, as listed below:

- Third-party damage
- Defects within the pipe during the rolling stage
- Damage incurred during pipe handling and installation
- Internal/external corrosion
- Cracks
- Dents
- Buckles
- Gouges

Repairs to the pipeline can be completed in many different ways. The method of repair is based upon the seriousness of the repair required and the economics associated with the repair.

The repair methods discussed below are for internal/external corrosion of a pipeline.

3.10.1 PIPELINE REPAIR STANDARD

When completing repairs on a pipeline, current codes and practices, as well as regulatory policies, must be adhered to. This section provides guidance on defect evaluation and describes the repair methods for the different types of pipeline anomalies. These are recognized standards when evaluating an anomaly for potential repair. These standards are:

- CSA Z662 (Latest Edition): Oil and Gas Pipeline Systems
- ASME B31G-1991 R2004: Manual for Determining the Remaining Strength of Corroded Pipelines. (This is a Supplement to ASME B31 Code for Pressure Piping.)
- API 1104: Standard for Welding Pipelines and Related Facilities
- ANSI/API RP 579: Fitness-for-Service (Recommended Practice)
- DNV-RP-F113: Pipeline Subsea Repair – Rules and Standards
- DNV-RP-F102: Pipeline Field Joint Coating and Field Repair of Line Pipe Coating
- ASME B31.8S: Managing System Integrity of Gas Pipelines

3.10.1.1 Corrosion

Corrosion can be generalized as a loss of pipe wall thickness and can be categorized as general surface corrosion, localized external pitting corrosion, localized internal pitting corrosion, or some combination of the above. When evaluating a corrosion anomaly, both of the above standards are to be considered.

Before evaluating the corrosion anomaly, the corroded areas shall be thoroughly cleaned, to remove corrosion products, so that their dimensions can be measured accurately. For anomalies with a maximum depth of 10% or less of the nominal pipe wall thickness, no repair is required.

For anomalies with a depth greater than 10% and less than 80% of the nominal wall thickness of the pipe, an evaluation is required to determine whether a repair is required. The anomaly shall be evaluated in accordance with the criteria in CSA Z662 (Latest Edition), Corrosion Imperfections. These criteria will calculate the maximum allowable longitudinal length of the corroded area. (ASME B31G-1991 R2004 contains the identical criteria to CSA Z662 (Latest Edition) for determining the maximum allowable longitudinal length of the corroded area.)

If the corroded area is greater than the maximum allowable longitudinal length, a safe maximum operating pressure (P) can be calculated using the ASME B31G-1991 R2004 standard. According to ASME B31G-1991 R2004, if the established MAOP of the pipeline is equal to or less than P, the corroded area may still be used for service at that MAOP. If the MAOP is greater than P, then either:

- A lower MAOP should be established such that it does not exceed P.
- The corroded area must be repaired or replaced.

3.10.1.1.1   Gouges, Grooves, and Arc Burns

All types of gouges, grooves, and arc burns shall be considered to be a defect as per CSA Z662 (Latest Edition), Gouges, Grooves, and Arc Burns. If a defect of this type is located within the pipe, it shall be removed by grinding until the ground area blends smoothly into the adjacent pipe. All repairs must be completed in accordance with CSA Z662 (Latest Edition), Grinding Repairs.

Dents

A dent is a depression in the pipeline caused by an external loading, producing a visible curvature/indentation to the wall of the pipe. A dent is considered smooth if it does not contain a stress concentrator (gouges, grooves, cracks, arc burns, etc.). Dents which exceed a depth of 6 mm in 101.6 mm OD pipe and smaller, or 6% of the outside diameter of pipe larger than 101.6 mm OD, shall be considered defects. Pipe containing such defects shall be repaired in accordance with CSA Z662 (Latest Edition), Dents.

In order to accommodate the smooth passage of internal inspection tools, dents which exceed prescribed limits of the tools shall be removed from the pipeline.

If a smooth dent is located on a mill or field weld and exceeds 6 mm in depth, it is considered a defect and must be repaired in accordance with CSA Z662 (Latest Edition), Dents. It is good practice to replace all dents that affect the curvature of the pipe at the seam or at any girth weld.

3.10.1.1.2   Dents Containing Stress Concentrators

All dents, which contain a stress concentrator, shall be considered a defect. The stress concentrator shall be removed by grinding until the ground area blends smoothly into the adjacent pipe. Pipe containing such defects shall be repaired in accordance with CSA Z662 (Latest Edition), Section 10.8.2.4 – Dents and CSA Z662 (Latest Edition), Grinding Repairs. If the defect is repaired by grinding, magnetic particle and ultrasonic inspections must be completed to verify that the crack has been completely removed.

3.10.1.1.3   Buckles

A buckle is defined as a localized distortion of the pipe wall, normally resulting from point loading on the pipe in a location of high bending stresses, and is characterized by the creasing of the pipe wall. There is a deflection in the pipe axis at the point of buckling.

Buckling to any degree is considered a defect and must be repaired. It is a good practice to replace any buckle, which is found within the pipeline.

3.10.1.1.4   Cracks (Non-Leaking)

All pipe body surface cracks are defects. Pipe containing such defects shall be repaired in accordance with CSA Z662 (Latest Edition), Pipe Body Surface Cracks, and CSA Z662 (Latest Edition), Grinding Repairs.

3.10.1.1.5   Replacement Pipe

The failed section of pipe shall be replaced with a section similar or equivalent to that of the existing pipe. The minimum length of replacement pipe shall be 2 m. The wall thickness of the replacement section shall be equal to or greater than that of the failed section.

The difference in wall thickness between the existing and replacement pipe shall not exceed 1.6 mm. The requirements of standards such as CSA Z662 shall also be met. Current applicable requirements can be found in Clauses 7.3 and 16.6.3 of CSA Z662.

Testing requirements for pipeline repairs are as follows:

- The replacement section shall be pre-tested to the same conditions as for a new pipeline:
  - For pipelines licensed for, or containing any amount, of H_2S pressure tested to 1.4 times the MOP.
  - For pipelines licensed as sweet and not containing any amount of H_2S, pressure tested to 1.25 times the MOP.

Pressure testing shall not be less than one hour in duration. A copy of the test chart and associated documentation (including a signed affidavit verifying the pressure test) shall be filled.

Codes such as CSA Z662 dictate additional requirements for pressure testing new pipelines, but these are not applicable to pre-testing replacement pipe.

- Complete the Pipeline Pressure Test Report and forward to the Director of Operations
- Visual inspection of the weld for pinholes, weld cap alignment, and excessive amount of weld material
- 100% radiography of all repair welds

3.10.2 REPAIR METHODS

Defects in pipelines may be repaired by a variety of methods, both temporary and permanent. The selection of the most appropriate repair method will depend on a number of factors such as feature size and location, accessibility, operating conditions, life criteria, and so forth. Those that have been commonly used by pipeline operators are highlighted below.

3.10.2.1 Dig and Replace

Excavate sections of the pipeline with corrosion defects. Remove the defects (cut the defect). Remove a defective section of the pipe and replace it with a new pipe. Cut the defective section and replace it with a pre-tested section of sound pipe; tie-in welds are inspected and returned to service. If the pipe is not pre-tested, then it has to be hydro-tested before the pipeline is returned to normal service. Removal and testing of the pipe section will necessitate shutdown and depressurization of the pipeline.

In some circumstances, the most cost-effective and/or safest way to repair a pipeline defect is to remove the affected segment of pipe and replace it. Removal requires that the pipeline be shut down or that the affected segment be isolated and depressurized. When the pipeline is shut down, it is cleaned and flushed with an inert gas to remove crude oil and reduce any explosive conditions. To isolate the affected segment without removing the product, operators can shut valves on either end of the defective segment. Operators sometimes use a freeze plug for the same purpose. A freeze plug is a procedure that uses liquid nitrogen to freeze the product before and after the segment that is to be replaced. Once the product has been removed or isolated, the pipeline is cut out as a cylinder and replaced by an already hydrostatically tested pipe to ensure it can withstand the operating pressure. After the tie-in welds are inspected, the product flow may resume. Operators may elect to remove a pipeline, while the product is still in the line using a method known as a hot tie-in. By maintaining a low positive pressure in the pipeline, specially trained personnel can weld and cut the pipeline without explosions while igniting the escaping product. Removal repair is one of the costlier repair options; however, it is considered a permanent fix for any pipeline defect (Table 3.28).

TABLE 3.28

Advantages and Disadvantages of Pipe Section Replacement (Roland and Dominic 2000)

Advantages	Disadvantages
Permanent	Expensive
Pipe may be re-designed to ensure problem does not re-occur, e.g. a change in the type of steel to one that is not susceptible to a particular corrosion mechanism	Takes a long time and the pipeline has to be taken out of service for some period, unless the tie-in can be hot tapped, further increasing expense
Can be used for extensive damage	

3.10.2.2   Composite Sleeve

Composite sleeves are used for repair of non-leaking pipeline defects. Most of the composites are fiberglass materials or carbon fiber-based composites. There are two basic types of fiberglass composites being used as reinforcement sleeves: rigid material (limited to relatively straight sections of pipe) and flexible material (can be applied to bends, elbows, and tees).

Composite sleeves usually consist of three parts:

a. A unidirectional composite wrap material
b. A two-part polymer adhesive between the wrap and the pipe and between layers of the wrap
c. A high compressive strength filler compound for load transfer

The advantages of composite reinforcements compared with steel sleeves are easier handling of the materials, lower skill requirements for installation personnel, more rapid installation, no shutdown required for installation, and lower overall cost.

On the other hand, the composite sleeve is not yet taken as a permanent fix to the pipeline. Similar to wraps, these sleeves may require replacement in the future (Table 3.29).

TABLE 3.29

Advantages and Disadvantages of Composite Sleeve Repair Method (Roland and Dominic 2000)

Advantages	Disadvantages
Strengthens the pipe in the area of the defect	Limited experience on through wall (leaking) defects
Proven in service and now widely accepted	Most systems will not provide axial strength
Permanent	
Simple	
By using multiple repairs relatively long lengths can be strengthened	

3.10.2.3   Full Encirclement Steel Sleeves

These sleeves are used widely for the repair of onshore pipelines. Because they involve welding, sleeves are not applicable for the repair of offshore pipes. Evaluate the fatigue resistance of sleeve before use. Use epoxy filler between the sleeve and carrier pipe.

a. Mechanically tightened or fiberglass reinforced sleeve (type A sleeve)
b. Compressive sleeve repair technique
c. Type B sleeve (pressure containing): used to repair a leaking defect and non-leaking defects

Steel sleeves are two halves of a cylinder of pipe that are placed around the defective segment of pipe. There are two types of steel sleeves: type A and type B. Type A sleeves reinforce pipe segments without being welded to the pipe itself. They are favored for their very simple construction; however, they cannot be used to repair circumferentially oriented or leaking defects because they do not reduce longitudinal stress or contain pressure. More recently, composite reinforcement sleeves have been used in a similar way to type A steel sleeves. Composite reinforcement sleeves are proprietary products and are most commonly made of fiberglass. They are approved for specific applications, such as blunt corrosion defects and blunt wall loss. Type B sleeves are welded to the carrier pipe at each end, so they can be used to repair leaks and strengthen circumferentially oriented defects. Because they are expected to contain leaks, type B sleeves must be designed to carry the full pressure of the carrier pipe (Tables 3.30 and 3.31).

TABLE 3.30

Advantages and Disadvantages of Type A Sleeve (Roland and Dominic 2000)

Advantages	Disadvantages
Proven in service	Will not contain a leak, therefore not suitable for active internal corrosion
Permanent	Requires welding, which is a significant disadvantage for subsea use
Simple	Difficult to inspect welds
Contractors are familiar with the application of welded sleeves, as they have been used in the pipeline industry for many years	Will provide some defect restraint but will not prevent defect failure
	No axial strength

TABLE 3.31

Advantages and Disadvantages of Using Type B Sleeve (Roland and Dominic 2000)

Advantages	Disadvantages
Will contain a leak	Requires welding, which is a significant disadvantage for subsea use
Proven in service	Difficult to inspect welds
Permanent	Fillet weld at sleeve end is often a source of defects
Simple	Requires welding to the pipeline, which can be difficult where the pipe wall is thin, or the product flow cools the wall rapidly
Contractors are familiar with the application of welded sleeves, as they have been used in the pipeline industry for many years	In the event of a leak the fluid may corrode the sleeve material
Limited axial strength	Will provide some defect restraint but will not prevent defect failure

3.10.2.4 Weld Deposition Repair

A pre-formed steel plate is welded to the pipeline over the defect. It is applied to the internal wall loss. It is used for tees and elbows (Table 3.32).

TABLE 3.32

Advantages and Disadvantages of Welded Patches (Roland and Dominic 2000)

Advantages	Disadvantages
Will contain a leak	May not prevent the failure of the defect
Permanent	Requires welding, which is a significant disadvantage for subsea use
Simple	Difficult to inspect welds
	Fillet weld around patch is often a source of defects
	Requires welding to the pipeline, which can be difficult where the pipe wall is thin, or the product flow cools the wall rapidly
	In the event of a leak the fluid may corrode the sleeve material
	Will provide some defect restraint, but will not prevent defect failure
	Is only practical for small areas

3.10.2.5   Grinding

Grinding is appropriate for internal pipe defects up to 0.4 t depth. It is a common method used to repair gouges and cracks in pipelines, but it is only effective under specific circumstances. When all the damaged metal can be removed and the pressure-carrying capacity is not reduced, grinding can be considered a permanent repair. Federal law in Canada and the United States limits the amount of pipeline that is ground out to 40% of nominal wall thickness. In situations where repairing a gouge or crack would require grinding more than 40% of nominal wall thickness away, operators can combine grinding with a steel reinforcement sleeve or composite.

3.10.2.6   Hot Tapping

This method removes defects from an in-service pipeline.

3.10.2.7   Wraps

Temporary wraps can be placed on the pipeline in the areas of the defects. These wraps are designed to provide additional corrosion protection and protect the existing defects.

The wrap can be applied in-situ without shutting down or depressurizing the pipeline; however, it is not a permanent repair for the existing defects. Wraps may not be able to withstand the pipeline MAOP and, therefore, would require replacement, with a more permanent method, if long-term fitness for service becomes an issue.

3.10.2.8   Clamping

Clamps can be used to repair a leaking external corrosion pit. The clamps normally have elastomeric seals to contain the pressure if the pipeline is leaking at the defect.

These clamps are designed to contain full pipeline pressure, so they are generally rather thick and heavy because of the large bolts used to provide the required clamping force. Breaking containment is not necessary for the installation of the clamps; therefore, no shut down of the pipeline would be required (Roland and Dominic 2000).

Mechanical clamps include bolt-on clamps and leak clamps. Bolt-on clamps are designed with elastomeric seals that can contain the full pipeline pressure. Operators may elect to weld bolt-on clamps to the pipeline to protect against failing seals. Bolt-on clamps can also be made with circumferential clamping mechanisms at each end of the sleeve. These designs make bolt-on clamps a permanent solution for leaking defects and circumferential cracks. Leak clamps are used to repair leaking external corrosion pits. These clamps have a sealing plug on the inside that can be screwed into a defect. These clamps can be made into a permanent repair if they are encapsulated by a domed fitting (Tables 3.33 and 3.34).

TABLE 3.33

Advantages and Disadvantages of Using Mechanical Clamps (Roland and Dominic 2000)

Advantages	Disadvantages
Will contain a leak	May not prevent the failure of the defect
May be made permanent	Generally temporary as rubber seals will degrade over time
Proven in service	Must be welded to the pipe to make a permanent repair
Relatively quick and simple to install	Heavy and difficult to handle
Some types can provide axial strength	In the event of a leak the fluid may corrode the clamp material
	Will only repair a relatively limited length

TABLE 3.34

Comparison of Repair Methods (Roland and Dominic 2000)

Repair methods	Permanent	Requires shut-in	Material cost	Ease of repair	Repair time frame	Repair cost
Wraps	X	X	Low	Easy	Short	Low
Composite sleeves	X	X	Low	Easy	Short	Low
Welded sleeves	✓	✓	Medium	Medium	Medium	High
Clamping	✓	X	Medium	Easy	Medium	Medium
Cut and replace	✓	✓	High	Difficult	Large	High

3.10.2.9   Dig and Recoat

1. Locate the site where a pipeline feature has been identified and stake the site using temporary markers.
2. Conduct an environmental/risk assessment, obtain applicable permits, and develop a comprehensive excavation plan.
3. Expose the pipe carefully using hand digging or hydro excavation with a vacuum truck. Hand digging assures the protection of the pipe from accidental damage, any mechanical excavation using a backhoe/excavator.
4. During the next step, the pipe coating is removed and the pipe is examined using non-destructive examination. Based on the field assessment results and regulatory requirements, repairs should be completed.
5. Recoat the pipeline with the appropriate coating as discussed in Sections 3.1.2 and 3.13. Upon completion, the site should be backfilled and restored to the original condition or better.

3.10.2.9.1   Internal Liners

The pipeline may be fitted with an internal liner that will prevent further corrosion. There are a variety of methods available to do this:

- Cured in-place – a thin reinforced textile liner, fixed by adhesive
- Modified slip lining – polyethylene liner compressed and expanded
- U-Process – polyethylene deformed to U shape, rolled in and expanded
- High-density Polyethylene – installation method not established (Table 3.35)

TABLE 3.35

Advantages and Disadvantages of Internal Liners (Roland and Dominic 2000)

Advantages	Disadvantages
Permanent	Pipeline must be taken out of service and cleaned
May not need to excavate	Can only be installed in short lengths (up to 800 m)
Possible to repair relatively extensive corrosion	Ends of the liner have to be sealed/joined to next section
	Concerns over use on sour service pipelines
	May interfere with internal inspection

3.10.2.10   Choosing the Most Appropriate Method of Repair

After it has been determined that a repair is required, the next question that must be answered is: "How are we going to repair it?" The answer to this question varies and is based on many different elements. There are many acceptable repair methods recognized by the industry; however, not all repair methods are applicable in all situations and some repair methods require special procedures to be in place in order to ensure they are properly completed. Timing, material availability, or the additional procedures that may have to be in place – all can have an impact on the method selected to repair a pipeline. Some repair methods just may not be practical, while others just may not be applicable or appropriate. Therefore, first it is important to consider the limitations of each repair method in order to determine whether the repair method is valid, per industry codes, standards, and best practices, and second whether it is feasible with respect to the specific situation. That being said, there are three repair methods recognized by industry standards which are acceptable for repairing any type of potentially injurious condition. Those repair methods are:

- Replace as cylinder
- Type B sleeve
- Mechanical bolt-on clamps

Below are some examples of items that must be considered prior to implementing other repair methods and why.

1. Does the defect, damage, or anomaly affect a seam or girth weld? If so, the repair methods available may be limited, depending on whether the weld is ductile or brittle. For example, the only methods applicable for repairing a 3% dent with 5% strain affecting a brittle weld in a gas pipeline are Replace as Cylinder, Type B Sleeve, or Mechanical Bolt-On Clamp. Whereas for the same dent affecting a ductile weld, there are additional repair options which are acceptable: Composite Sleeve, Type A Sleeve, or Hot Tap.
2. Is the defect, damage, or anomaly leaking? If so, the only repair methods allowed are Replace as Cylinder, Type B Sleeve, Mechanical Bolt-on Clamps, and, depending on the specific scenario, Removal by Hot Tap.
3. Can the pipeline be shut down or diverted? If not, then this eliminates the option to Replace as Cylinder.
4. What is the size of the targeted repair? Defects which cannot be contained entirely within an NPS 3 fitting cannot be repaired using a fitting type repair method. Additionally, defects which exceed the largest possible coupon of material that can be removed through a hot-tap fitting cannot be repaired by using a hot tap to remove the defect, damage, or anomaly from the pipeline.
5. Schedule/availability of materials: How soon is the repair required to be completed? Some repair methods such as using mechanical bolt-on clamps are designed for a single specific application and may have longer lead times to obtain the repair clamp, which eliminates it as a repair option depending on the repair schedule.
6. Are all required procedures and qualified personnel available to perform the repair? Regulations require that some repairs, such as deposition of weld metal, be performed by a qualified welder using an approved welding procedure. If these required items are not already in place, they can cause delays to the repair schedule.

4 Construction and Materials Defect

4.1 QUALITY CONTROL PROGRAM

With every project comes the possible exposure to construction defect losses. Having a quality assurance/quality control (QA/QC) program, however, can help mitigate construction and material defects. Three program elements to consider are:

- QA/QC program administration and documentation
- A subcontractor selection procedure
- Contractual risk transfer (CRT)

Many oil and gas companies have a QA/QC program in place; it is good only if it is enforced. Keeping the QA/QC program updated and documented, selecting the right subcontractors, having contractual risk transfer implemented, and enforcing accountability for the use of the program can help make the difference between a quality project and a problematic one, possibly damaging your hard-earned repute (Lawrie 2018).

4.1.1 QA/QC PROGRAM ADMINISTRATION AND DOCUMENTATION

A number of factors comprise program administration and documentation, such as management commitment and responsibility, design codes, and documentation (Lawrie 2018).

4.1.1.1 Management Commitment and Responsibility

Your quality control program should start with management commitment and include responsibility for program administration and implementation. Some considerations include:

- A written QA/QC program, signed by the head of your company
- A plan and budget to implement/support the program
- Individuals with QA/QC responsibilities and accountability
- A formal commissioning program with the project owner

4.1.1.2 Materials

Material selection, storage, and handling should include inspection, evaluation, and documentation/retention, confirming that materials comply with codes, design specifications, and national standards (Lawrie 2018). This should be done both before and after installation. A few additional considerations include:

- Understanding new technologies, materials, and work methods
- Understanding manufacturers' warnings and instructions
- An acceptance inspection by the supplier, if available
- Notification process to the supplier if defects are identified

4.1.1.3   Design Codes and Standards

A complete and final approved set of current contract drawings and specifications, including design changes, should be in hand before work begins. Be sure to include them.

Efforts should be made to enforce strict quality control standards such as ISO 9000.

4.1.1.4   Quality Inspections

Construction plans should be reviewed, tested, approved, and documented. The person inspecting (you, the subcontractor, or a third-party inspector) must be qualified to make the inspection and should do so at several phases, including:

- Prior to the start of a project or task
- During the construction process
- When there are changes in the project, process, or materials
- When the project is complete
- Upon warranty expiration

4.1.1.5   Workmanship

Using the most recently approved design documents, product/manufacturer instructions, and specified and approved materials can help enhance the quality of the workmanship. Do not substitute materials unless the substitutions are approved by the design team in writing (Lawrie 2018).

4.1.1.6   Documentation and Retention of Documentation

Many types of documents are produced in the course of a project. Thorough documentation, including verbal changes and retention of the documents, can play an important role in the event of a construction defect loss, which may occur years later. *A common mistake is not fully documenting verbal changes made in the field.*

4.1.2   SUBCONTRACTOR EVALUATION AND SELECTION

Subcontractors play a significant role in the workmanship and quality outcome of your project. To help select quality workers for your project, start by having a subcontractor evaluation and selection procedure and program in place. Such a program could include:

- Checking their financials, references, project experiences, and losses
- Requiring appropriate licenses and certifications to perform the work
- Evaluating the subcontractor's in-house safety and QA/QC written program/ procedures

The following subsection gives examples of some requirements for inspectors.

4.1.2.1   Contractor and Consultant Competency and Procurement

Pre-qualify all integrity contractors and suppliers, and engage only those who are approved. All contractors shall meet the requirements of the Company Contractor Evaluation Criteria and the Public Procurement Act of the project country. In addition, all integrity contractors and suppliers shall also have the appropriate technical requirements, as stated below.

4.1.2.1.1   Corrosion Chemical Providers

1. The chemical vendor field representatives shall possess extensive experience in the selection, application, and performance review of corrosion inhibitors.
2. They shall qualify their selected chemicals for the applicable operating conditions via laboratory testing.

4.1.2.1.2   Cathodic Protection Service Providers

1. All annual survey data shall be collected by a NACE CP Level I or higher tester or a person possessing equivalent industry experience.
2. All reports and recommendations shall be reviewed and signed by a registered professional engineer, a certified engineering technologist, or, at minimum, a NACE CP Level II Technician.

4.1.2.1.3   Non-Destructive Testing Vendors

1. The field evaluator shall have at least a Level II designation in the applicable NDT field (radiography or ultrasonic).

4.1.2.1.4   In-Line Inspection Vendors

The vendors shall:

1. Provide detailed, safe work procedures on in-line inspection tool preparation, launching, and retrieval
2. Demonstrate a sound knowledge of defect assessment methodology (such as determining the remaining strength of a corroded pipe)
3. Ensure the final in-line inspection report is reviewed and signed by a data analyst with no less than five years of experience in in-line inspection data analysis and reporting
4. Identify and report all immediate repair features within 45 days of completion of the inspection

4.1.3 CONTRACTUAL RISK TRANSFER

Contractual risk transfer is a program designed to help protect you against potential liability for the actions, services, or products of other companies/persons associated with your project. A few considerations include:

- A written contract of requirements, responsibilities, and accountabilities
- Insurance coverage specifications (auto, general liability, worker's compensation, etc.)
  - Insurance – contractors are required to provide a copy of their liability insurance.
  - Worker's compensation – contractors are required to provide a copy of their worker's compensation coverage and their injury rating.
- Indemnification agreements
- Current certifications of insurance and additional insured endorsements

4.2 QUALITY ASSURANCE OF PIPELINE DESIGN

The overall design of a pipeline will be determined by its location, the type of fluid being carried, and its operating pressure and temperature.

The operating pressure will determine the grade and the wall thickness of the materials, and the temperature will affect the requirements for pipe coatings, insulation, expansion joints, anchor blocks, etc.

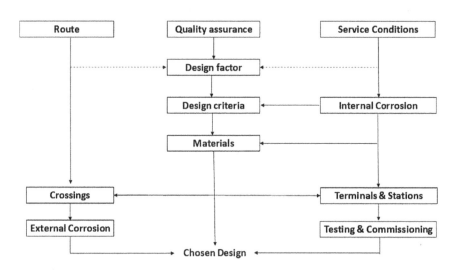

FIGURE 4.1 Sequence of pipeline design (Source: Kadir, Ali. 2002. *Distribution, transmission system and design*. Lecture notes, Salford: School of Computing Science & Engineering – University of Salford).

Gas pipelines require special attention over the above normal design requirements for liquids because of the vast quantity of stored energy in the pipeline steel due to the compressibility of gas. The design should therefore take into account the probable consequences of failure and the parameters that can be adjusted to minimize the possibility of failure.

In terms of quality assurance (see Figure 4.1), the following factors must be considered in any pipeline design project:

- Design factor
- Design criteria
- Materials

4.2.1 DESIGN FACTOR

This is usually expressed as a percentage of the specified minimum yield strength (SMYS) for the pipeline steel. It has been shown that the boundary between arrest and propagation in brittle fractures is dependent on the operating temperature and the energy stored within the pipe material, i.e. the hoop stress. Therefore, it would seem appropriate to specify the design factor at a level which takes into consideration the boundary between defect arrest and propagation, also known as the leak/break boundary. Obviously not all defects behave in the same way; their behavior depends on the length, depth, and size of the pipe, and therefore some method of correlating these parameters is required. Figure 4.2 shows the results of a large number of tests carried out by British Gas. The results are plotted as the percentage of SMYS at failure against a defect length factor (which does include depth) for both corrosion defects and artificially induced defects. The leak/break boundary is clearly evident in each case and shows that the mechanically induced defects fail at a lower stress level than corrosion defects; these are therefore the controlling influence on design. It should also be noted that there are no breaks below 30% SMYS, and therefore a design factor of 0.3 would ensure a break-free pipeline under most circumstances.

Where pipelines are laid in open country, the likelihood of mechanical defects being introduced is very much reduced, and therefore it is appropriate to use a design factor higher than 0.3. IGE/TD/1 and ANSI B31.8 specify 0.8 as the maximum permissible design factor based on the following:

- The results in Figure 4.2 indicate that 80% is equivalent to a defect length factor of 2.0 and the likelihood of a defect of this size in open country is low.
- Pipelines have been operating for many years at this level without serious incidents.
- The minimum design and operating conditions specified for 80% SMYS operation have been demonstrated to provide an acceptable level of risk.

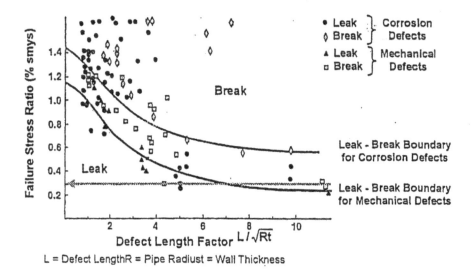

L = Defect Length R = Pipe Radius t = Wall Thickness

FIGURE 4.2   Defect length factor (Source: Kadir, Ali. 2002. *Distribution, transmission system and design.* Lecture notes, Salford: School of Computing Science & Engineering – University of Salford).

The two important design factors therefore are:

- 0.3 for pipelines operating in locations where a line break is not acceptable
- 0.8 for pipelines operating in an open country where a line break is unlikely but could be tolerated

ANSI B31.8 allows design factors up to 0.80 in very specific remote areas, and IGE/TD/1 allows design factors higher than 0.72 if these can be justified by risk analysis.

The design factor is different for liquid and gas designs. The liquid pipeline design for hydrocarbons generally utilizes a single design factor which does not vary by location or proximity. The gas pipeline design utilizes experience and experimental data to provide more mechanical strength for a pipeline in areas where either the possibility or consequence of any leak or rupture is more severe. Table 4.1 lists the liquid design factors and the maximum design factors available from each design code for gas pipelines (Kadir 2002).

TABLE 4.1
Design Factors for Liquid and Gas Pipelines

	ASME B31.8	ASME B31.4	PD 8010	IGE/TD/1	ISO 13623	EN 1594
Design factor (liquid)	–	0.72	0.72	–	0.77	–
Design factor (gas)	0.4–0.8	–	0.3–0.72	0.3–0.8	0.45–0.83	0.72

4.2.2 DESIGN CRITERIA

4.2.2.1 Stresses

4.2.2.1.1 Hoop and Radial Stress

Hoop stress, σ_H, is the stress in a pipe with a wall thickness, t, acting circumferentially in a plane perpendicular to the longitudinal axis of the pipe, produced by the pressure, P, of the fluid in a pipe with diameter, D, and is determined by Barlow's formula.

$$\sigma_H = \frac{PD}{2t} \tag{4.1}$$

where

σ_H = hoop stress, MPa
P = internal design pressure (gauge), MPa
t = pipe wall thickness, m
D = pipe diameter, m

4.2.2.1.1.1 Thick Cylinders   The full Lame equations are simplified for the design of thick-wall pipelines. A pipeline with $D/t < 20$ is known as a thick-wall pipeline. Take into consideration a thick-wall pipe, subject to internal pressure, P_i, with zero external pressure.

$$\text{Hoop stress, } \sigma_H = A + \frac{B}{r^2} \tag{4.2}$$

$$\text{Radial stress, } \sigma_r = A - \frac{B}{r^2} \tag{4.3}$$

The two well-known conditions of stress which allow the determination of the Lame constants A and B are:

$$\text{At } r = R_1 \quad \sigma_r = P_i$$

and

$$\text{at } r = R_2 \quad \sigma_r = 0$$

$$\text{Lamé constant } A = -P_i \left(\frac{R_1^2}{R_1^2 - R_2^2} \right) \tag{4.4}$$

$$\text{Lamé constant } B = P_i \left(\frac{R_1^2 R_2^2}{R_1^2 - R_2^2} \right) \tag{4.5}$$

Therefore, from Equations (4.2) to (4.5), the radial and hoop stress is calculated as:

$$\sigma_r = -P_i \left(\frac{R_1^2}{R_2^2 - R_1^2} \right) \left(1 - \frac{R_2^2}{r^2} \right) \tag{4.6}$$

$$\sigma_H = -P_i \left(\frac{R_1^2}{R_2^2 - R_1^2} \right) \left(1 + \frac{R_2^2}{r^2} \right) \tag{4.7}$$

The maximum radial and hoop (circumferential) stresses occur at $r = R_1$, when $\sigma_r = P_i$. The negative sign indicates tension.

$$\therefore \ \sigma_H = -P_i\left(\frac{R_1^2 + R_2^2}{R_2^2 - R_1^2}\right) \tag{4.8}$$

$$\sigma_r = P_i\left(\frac{R_1^2}{R_2^2 - R_1^2}\right) \tag{4.9}$$

where

σ_H = hoop stress, MPa
σ_r = radial stress, MPa
R_1 = internal radius, m
R_2 = external radius, m
r = radius at the point of interest (measured from the pipeline center)
P_e = external pressure (gauge), MPa
P_i = internal pressure (gauge), MPa

4.2.2.1.1.2 Thin-Wall Pipeline

A pipeline with $D/t > 20$ is known as a thin-wall pipeline. A basic approach is to use the thin-wall hoop stress theory. Since the maximum hoop stress is normally the limiting factor, it is this stress which will be considered.

It is predictably accurate for $D/t > 20$. The hoop stress is then calculated as below:

$$\sigma_H = (P_i - P_e)\frac{D_2}{2t}, \quad D/t > 20 \tag{4.10}$$

Hoop stress developed in the pipe wall at the internal design pressure is given by:

$$\sigma_H = (P_i)\frac{D_2}{2t}, \quad D/t > 20, \ P_e = 0 \tag{4.11}$$

where

σ_H = hoop stress, MPa
P_i = internal design pressure (gauge), MPa
P_e = external pressure (gauge), MPa
t = design thickness, m
D_1 = inside pipe diameter, m
D_2 = outside pipe diameter, m

4.2.2.1.2 Longitudinal Stress

The estimation of the longitudinal stress in a section of the pipeline requires the individual stress components to be identified by knowing external restraining conditions.

The axial (longitudinal) stress in a pipeline depends wholly on the limiting conditions (imposed boundary condition) experienced by the pipeline, i.e. whether the pipeline is unrestrained, restrained, or partially restrained. The boundary conditions can include the effects of soil reaction loads, anchor restraints, line pipe bend resistance, and residual pipe-lay tension forces. The longitudinal stress in a thin cylindrical shell is calculated as half of the hoop stress.

The total longitudinal stress should be the sum of the stresses arising from the following:

- Pressure
- Bending
- Temperature
- Weight
- Other sustained loads
- Occasional loadings

A pipeline should be considered totally restrained when axial movement and bending resulting from temperature or pressure change are totally prevented.

4.2.2.1.2.1 Fully Restrained Pipeline   A fully end-constrained boundary condition can occur at an anchor block, pig trap, pipeline end manifold (PLEM) or pipeline end termination (PLET) sled. For a fully end-constrained pipeline, the longitudinal strain ($\varepsilon_1 = 0$) and deflection ($\Delta = 0$) components are zero and the longitudinal stress response can be determined from Equation (4.12), assuming a constant uniform temperature field.

Piping in which the soil or supports prevent the axial displacement of flexure at bends is restrained. Restrained piping may include the following:

- Straight sections of buried piping
- Bends and adjacent piping buried in stiff or consolidate soil
- Sections of above-ground piping on rigid supports

The net longitudinal compressive stress in a restrained pipe is calculated based on Clause 419.6.4 of ASME B31.4 as

$$\sigma_{LR} = E\alpha\left(T_2 - T_1\right) - \upsilon\sigma_H \qquad (4.12)$$

where

σ_{LR} = restrained longitudinal stress, MPa
α = linear thermal coefficient of expansion, mm/mm/°C
T_2 = operating temperature (maximum or minimum metal temp), °C
T_1 = installation temperature, °C
υ = poissons ratio ($\upsilon = 0.30$ for steel)
σ_H = hoop stress, MPa
E = modulus of elasticity, GPa

Support and Anchors
Pipe supports should be used to prevent forces and moments being transmitted to connected equipment. Pipe supports should not cause additional local stresses. Avoid

welding supports or other attachments to the pipe. Anchor blocks are used to prevent the axial movement of a pipeline. Anchor block design should take into account any pipeline expansion force and any pipe-to-soil friction resisting movement.

Prevent expansion or contraction causing excessive force or stresses in pipeline components. Use pipeline anchors where necessary. Allow for buckling forces in areas of ground movement. Allow for buckling forces in areas of ground movement. Allow for the effect of constraints such as lateral branch connections, friction, etc., allow for stress intensification factors in components. Do not introduce stress intensification points in the pipe by using non-standard procedures. Calculations of thermal expansion/stresses should use the total operating temperature difference between the maximum and minimum metal temperatures. Forces which produce shear stress should be minimized.

Forces on an Anchor Block

Anchor blocks are used to stop the axial movement of a pipeline. Anchors are normally required when the pipeline comes above ground, prior to pig hatches, other branches, or manifolds. The pipeline is normally connected to the anchor block by the addition of slip-on flanges to the pipeline, which are fillet-welded in place. A large concrete block is then constructed around the pipeline to resist the expansion forces. Restrained portions are always prevented from moving by installing anchors and guides, but in a buried line a large portion is fully restrained only by soil friction.

The axial compressive force required to restrain a pipeline can be calculated as follows:

Thin wall

$$F = A\left(E \cdot \alpha\left(T_2 - T_1\right) + 0.5S_h - \upsilon S_h\right) \tag{4.13}$$

Thick wall

$$F = A\left(E \cdot \alpha\left(T_2 - T_1\right) + \frac{S_h}{k^2 + 1} - \upsilon\left(S_h - \frac{P}{10}\right)\right) \tag{4.14}$$

where

F = axial force, N
E = modulus of elasticity, GPa
T_1 = installation temperature, °C
S_h = hoop stress, MPa
υ = Poisson ratio (0.3 for steel)
A = cross-sectional area of the pipe wall, m^2
α = coefficient of thermal expansion, per °C
T_2 = maximum or minimum metal temperature, °C
k = ratio of the outside diameter to the inside diameter
P = internal design pressure (gauge), MPa

When designing with natural flexibility (accommodating expansion in the piping system), structural members can act as an anchor and support points for the system. Also, Clause 321.1.4 of ASME B31.3 may provide further information on pipe support materials.

4.2.2.1.3   Shear Stress
Shear stress in a pipeline should be minimized. The shear stress should be calculated from the torque and shear force applied to the pipeline, using the following equations:

$$\tau = \frac{1000T}{2Z} + \frac{2F_s}{A}$$ (4.15)

where

τ = shear stress, N/m^2
T = torque applied to the pipeline, Nm
F_s = shear force applied to the pipeline, N
A = cross-sectional area of the pipe, m^2
Z = section modulus of the pipe, m^3

Sections 4.1.2.1.3.1 and 4.1.2.1.3.2 show how to determine shear stress due to torsion and spanning, which makes up the total shear stress.

4.2.2.1.3.1   Shear Stress due to Torsion   Torsion is the twisting of a straight bar when it is loaded by twisting moments or torques that tend to produce rotation about the longitudinal axes of the bar. When subjected to torsion, every cross-section of a circular shaft remains plane and undistorted, and then the bar is said to be under pure torsion.

The shear stress on a uniform cylindrical shaft which is under a uniform torsion is given by:

$$\tau = \frac{Tr_x}{J} = \frac{Tr_x}{2I_x} = \frac{T}{2Z}$$ (4.16)

Thin-Walled Tube
For a thin cylindrical shaft with $t < R/10$,

$$J = R^2 \left(2\pi rt \right) = 2\pi R^3 t = \frac{\pi}{4} D^3 t$$ (4.17)

$$\tau_{max} = \frac{TD}{2\left(\frac{\pi}{4} D^3 t \right)} = \frac{2T}{\pi D^2 t}$$ (4.18)

The maximum shear stress due to torsion can be calculated as follows:

$$\tau_{max} = \frac{TD_o}{2J} = \frac{TD_o}{4I_x} = \frac{16TD_o}{\pi \left(D_o^4 - D_1^4 \right)}$$ (4.19)

where

τ = shear stress, N/m^2
T = torque or twisting moment, Nm
R = radial distance from the longitudinal axis, m

I_x, I_y = moment of inertia about the x and y axis, kg m^3
Z = section modulus, m^3
J = polar moment of inertia

4.2.2.1.3.2   Shear Stress due to Spanning   The shear stress due to spanning is composed of a vertical shear and a longitudinal shear due to the bending. The maximum shear force acting on a simple span is equal to the maximum support reaction. This is, in turn, equal to the change in shear force at the reaction. The maximum vertical shear stress is defined as the force per unit area. The maximum vertical shear stress is calculated as follows:

$$\tau_{max} = \frac{R_{max}}{A_s} = \frac{2F_s}{A_s}$$ (4.20)

where
F_s = shear force applied to the pipeline, N
A_s = cross-sectional area of the pipe, m^2
R_{max} = maximum vertical reaction on the pipe, N

4.2.2.1.4   Maximum Allowable Stress

Pressure and temperature, as well as other operating conditions such as bending, can create expansion and flexibility problems, and therefore stress criteria are specified in all codes, limiting the level of combined stress allowed in a pipeline. The design factor relates only to hoop stress; if other stresses are significant, then these could contribute to the increase in the yield stress of the pipeline steel. Design codes vary in the way they calculate the combined or equivalent stress, but the following equation is typical:

$$S_e = \sqrt{\left(S_h^2 + S_l^2 - S_h.S_l + 3\tau_\tau^2\right)}$$ (4.21)

where
S_e = equivalent stress (N/mm^2)
S_L = longitudinal stress (N/mm^2)
S_h = hoop stress (N/mm^2)

The maximum allowable equivalent stress in all codes is 90% of the SMYS.

4.2.2.1.4.1   Allowable Equivalent Stress

$$\sigma_{ae} \leq FS_y$$ (4.22)

In accordance with Clause 833.4 of ASME B31.4, the maximum allowable equivalent stress is 90% of the SMYS.

$$\sigma_{ae} = 0.9S_y$$ (4.23)

where
σ_{ae} = allowable equivalent stress
S_y = specified minimum yield strength

4.2.2.1.5   Allowable Hoop Stress
The allowable hoop stress may be calculated using any of the equations below (refer to Clause 805.234 of ASME B31.8 and Clause 201.4.1 of API RP 1111).

$$\sigma_{aH} = f \times e \times T \times S_y \tag{4.24}$$

where

σ_{aH} = allowable hoop stress
S_y = specified minimum yield strength
f = design factor
e = weld joint factor
T = temperature derating factor

The allowable hoop stress for a cold worked pipe is 75% of the above value (Clause 201.4.4 of API RP 1111). The design factor is 0.72 for pipelines and liquid risers, 0.60 for gas risers, and 0.50 for gas platform piping.

4.2.2.2   Fatigue Life

Pressure cycling (i.e. hoop stress) can cause small weld defects to grow in time to a critical size, and it is therefore the major factor in determining the fatigue life of welded steel gas pipelines, particularly those pipelines which are designed for use as line-pack storage. Fatigue life is not greatly influenced by temperature, provided that the fracture toughness properties are met, because temperature effects are small in comparison with those produced by hoop stress. When a new pipeline is hydrostatically tested at a high level, any existing defects will grow under the influence of the high-stress level and any which reach a critical length will fail, resulting in pipe rupture. Normally pipelines do not fail, and any remaining defects are therefore noncritical. Pressure cycling, however, causes a gradual growth in the remaining defects such that one or more could become critical in time. Restrictions on pressure cycling are required to prevent this, particularly on pipelines operating under line-pack conditions. British gas has determined that a pipeline life of 40 years, assuming 1 cycle/day, is equivalent to 15,000 cycles of 125 N/mm² magnitude.

The maximum permissible fatigue life in IGE TD/1 has therefore been set as 15,000 cycles at 125 N/mm². If any cycles are less than 125 N/mm², they are multiplied by a factor based on the number of stress cycles at the lower level required to cause the same damage as 15,000 cycles at 125 N/mm².

It is important, therefore, that:

- The design life of the pipeline is known and designed for at the start.
- An accurate log of pressures and cycles is kept throughout the life of the pipeline.

From the pipeline log, it is possible to determine the remaining life. If this is less than the required life, then:

- The operating mode can be adjusted accordingly.

- The pipeline can be subjected to a high-level test before 15,000 cycles are reached. A successful high-level test would ensure further 15,000 cycles of fatigue life.
- An in-line crack detection tool can be used to determine the condition of the pipeline. The revalidation period of the fatigue life will depend on the resolution/detection limits of the tool used.

Figure 4.3 shows the results of fatigue life tests carried out by British Gas on pipes operating at a maximum of 30% SMYS. A daily stress range of 125 N/mm² (18,000 ibf/in²) is equivalent to a pressure cycle of 530 ibf/in² in a 24-inch diameter pipeline of grade 5LX60 steel with a wall thickness of 0.0375 inches. If this range of stress cycling was repeated each day, the pipeline would have a fatigue life of 40 years; however, this would be unusual for UK pipelines and the fatigue life would probably be well in excess of 40 years.

Minor defects in the pipe material or welds can grow under the influence of stress cycling. One stress cycle per day of 125 N/mm² could lead to failure after 40 years (15,000 cycles). This can be important for gas pipelines operating under line-pack conditions. The number of stress cycles expected during the life of a pipeline should be calculated. If this approaches 15,000 then the pipe material could be changed or a record of daily stress cycles should be maintained over the life of the pipeline. If the number of stress cycles reaches 15,000, then the pipeline should be revalidated by hydrostatic testing.

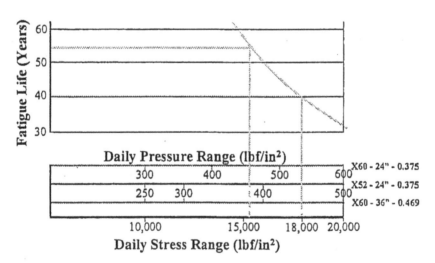

FIGURE 4.3 Fatigue life test results (Source: Kadir, Ali. 2002. *Distribution, transmission system and design*. Lecture notes, Salford: School of Computing Science & Engineering – University of Salford).

4.2.2.2.1 Fatigue Limit

The significance of the fatigue limit is that if the material is loaded below this stress, then it will not fail, regardless of the number of times it is loaded. In accordance with

IGE TD/1, 15,000 cycles at 125 N/mm² have been set as the maximum permissible fatigue life.

4.2.2.2.2 S–N Curve

A very useful way to visualize the time to failure for a specific material is with the S–N curve. This is a graph of the magnitude of cyclic stress (S) against the logarithmic scale of cycles to failure (N).

A fatigue life test should be conducted for the pipe material, and an S–N curve should be drawn. For instance, a specimen of the pipe material is placed in a fatigue testing machine and repeatedly loaded to a certain stress, σ_1. The loading cycles are continued until failure occurs, and the number, n, of loading cycles to failure is noted. Then the test is repeated for a different stress, say σ_2; if σ_2 is larger than σ_1, the number of cycles to failure will be less. If it is smaller, the number of cycles to failure will be more.

Eventually, enough data are accumulated to plot an S–N curve. Such curves have the general shape shown in Figure 4.4, where the vertical axis is usually a linear scale and the horizontal axis is a log scale.

From Figure 4.4, the smaller the stress, the larger the number of cycles to produce failure. Fatigue strength curves ("S–N" curves) for a particular material or structural weldment formation can be found in design standards such as ASME boiler and pressure vessel codes, Section VIII, Division 2, Appendix 5, and BS 7910, etc.

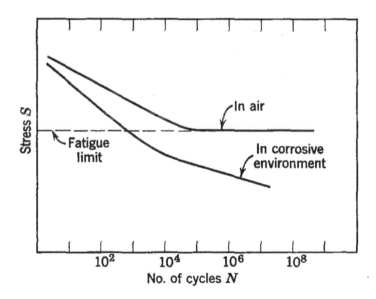

FIGURE 4.4 Typical S–N curves for steel.

4.2.2.3 Expansion and Flexibility

Avoid expansion bends and design the entire pipeline to take care of its own expansion. Maximum flexibility is obtained by placing supports and anchors so that they will not interfere with the natural movement of the pipe. Allow for stress intensification factors in components.

Expansion joints may be used to avoid pipeline bending (flexure) stress due to the movement of supports or the tendency of the pipe to expand under temperature change. The following should be considered for the use of expansion joints.

- Select the expansion joint carefully for the maximum temperature range (and deflection) expected so as to prevent damage to expansion fitting.
- Provide guides to limit the movement at the expansion joint to direction permitted by the joint.
- Provide adequate anchors at one end of each straight section or along its middle length, forcing the movement to occur at the expansion joint, yet providing adequate support for the pipeline.
- Mount expansion joints adjacent to an anchor point to prevent the sagging of the pipeline under its own weight and do not depend upon the expansion joint for stiffness (it is intended to be flexible).
- Give consideration to the effects of corrosion, since the corrugated character of expansion joints makes cleaning difficult.

Formal flexibility analysis for an unrestrained piping system may not be required.

4.2.2.3.1 Flexibility and Stress Intensification Factors

The stress intensification factor (SIF) is defined as the ratio of the maximum stress state (stress intensity) to the nominal stress, calculated by the ordinary formulas of mechanics.

In piping design, this factor is applied to welds, fittings, branch connections, and other piping components where stress concentrations and possible fatigue failure might occur. Usually, experimental methods are used to determine these factors.

The structural analysis of risers, expansion loops, or tee assemblies entails the use of flexibility stress intensification factors applicable to accurately model the structural behavior of bends and tees within the system.

It is recognized that some of the SIFs for the same components are different for different codes. In some cases, different editions of the same code provide different SIFs for a given component. The way that the SIFs are applied to moment loadings is also different for different codes. The B31.1 and ASME Section III codes require that the same SIF be applied to all the three-directional moments, while the B31.3, B31.4, B31.5, and B31.8 codes require that different SIFs be applied to the in-plane and out-of-plane moments, with no SIF required for torsion.

Therefore, the stress analyst has to ensure that the appropriate SIFs from the applicable code are used.

Flexibility is added in a pipe system by changes in the run direction (offsets, bends, and loops) or by the use of expansion joints or flexible couplings of the slip joint, ball joint, or bellow type. In addition more or less flexibility can be added by changing the spacing of pipe supports and their function (e.g. removal of a guide close to a bend to add flexibility).

Another way to increase the flexibility is to change the existing piping material to a material with a higher yield or tensile strength or to a material quality that does not need additional corrosion and erosion allowance, and thereby obtaining a reduction in the wall thickness which again gives more flexibility since the moment of inertia is reduced with a reduction of the pipe wall thickness. The flexibility factors and stress intensification factors that can be used are listed in standards, e.g. ASME B31.8, table E1 and CSA Z662-11, table 4.8.

4.2.3 Materials

Line pipe material specification was until recently limited to API 5L. This specification provided the building blocks but was normally supplemented by additional company standards or specific project ones. In conjunction with the EN and ISO pipeline design codes, ISO line pipe specifications, ISO 3183, have also been issued, which have been incorporated into the EN system as EN 10208. The ISO and EN codes require the use of their own line pipe specifications and requirements when the code is used (as well as a number of other "normative" codes listed in each code).

The ISO/EN line pipe specifications are split into three categories:

- Class A
- Class B
- Class C

Class A is essentially a re-write of API 5L; class B adds some general amendments varying in such items as the under tolerance and offering numerous additional requirements subject to choice by the designers, and class C is for sour service and special pipes. Class B is the one most anticipated to be used for pipelines. The specifications also round up the SMYS in terms of kN/m^2 in which the standard API SMYS ratings vary by a few percent. The standard under tolerance for wall thickness for class B line pipe is reduced to 5% from 10% to 12.5% on the basic API 5L specification.

Line pipe grade is another area where changes in the standard have occurred in recent years. The most common grade of line pipes seen is now X60 or X65, certainly for larger-diameter and higher-pressure pipelines. Pipelines using X70, X80, and X100 are now being assessed or built. A minimum ratio of 90% of ultimate tensile strength (UTS) to SMYS is commonly applied.

4.2.3.1 Wall Thickness

Pipeline steels have standard values of nominal wall thickness, with tolerances to allow for the manufacturing process. Acceptable tolerances vary depending on the requirements of the pipeline owner, but a typical tolerance for under thickness is 5% for submerged arc welded (SAW) pipe.

The nominal wall thickness therefore is not necessarily the actual wall thickness. In ANSI B31.8 pipelines can be designed using the nominal wall thickness, but BS 8010 and IGE/TD/1 require the pipelines to be designed on the basis of "minimum wall thickness", i.e. the nominal wall thickness less the maximum tolerance for under thickness.

The minimum wall thickness should be equal to or greater than the design thickness t.

$$t = \frac{P \times d}{2 \times f \times s} \tag{4.25}$$

where

t = design thickness
d = pipe diameter
P = internal pressure
S = specified minimum yield stress of the pipe material
f = design factor less than 1.0

From the above equation it is clear that the design thickness is dependent on other parameters which will be determined by the required operating conditions. The pressure and diameter will be determined by the required transmission capacity, and the design factor will be dependent on the pipeline route; therefore, it remains to select a material grade which will give a suitable design thickness or vice versa.

An important consideration in the selection of pipe wall thickness, particularly for gas pipelines, is to know what wall thickness is required to resist penetration by mechanical equipment and what depth of defect will result from the different types of machinery likely to be encountered.

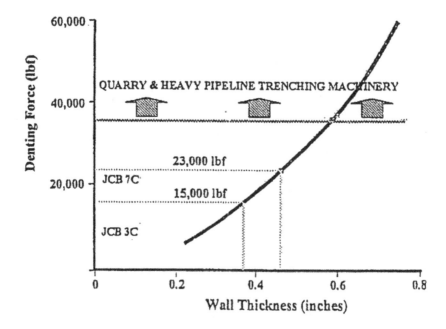

FIGURE 4.5 Test on X60 pipeline steel of different wall thickness (Source: Kadir, Ali. 2002. *Distribution, transmission system and design.* Lecture notes, Salford: School of Computing Science & Engineering – University of Salford).

Figure 4.5 shows the results of tests carried out by British Gas to find the force required to produce a 0.51 mm (0.2 inches) dent in X60 pipeline steel of different wall thicknesses. It will be observed that a wall thickness of 12.7 mm (0.5 inches) will resist impact by the lighter types of machinery encountered in quarrying and pipeline trenching; however, during normal operations it is unlikely that such equipment will be in use without the pipeline operator being aware of it. A wall thickness of 12.7 mm (0.5 inches) will resist an impact of 15,000 lbf, which makes it suitable for use in the more remote areas. The choice of the material grade should therefore reflect the requirement for a wall thickness appropriate to the risk category of the pipeline route. Additionally, the requirement for a thicker-walled pipe should not be taken as an opportunity to reduce the material grade in any area below that being used on the remainder of the pipeline. High-risk areas require a combination of appropriate wall thickness and grade of steel.

4.2.3.2 Material Properties

The pipe material must have sufficient fracture toughness properties to prevent propagating brittle or ductile fractures at the minimum operating temperature of the pipeline. In the United Kingdom, the normal operating temperature is taken as 5°C and it is required by IGE/TD/1 that fracture toughness properties are demonstrated at 0°C to give a margin of safety. In the case of brittle fracture, the requirement is

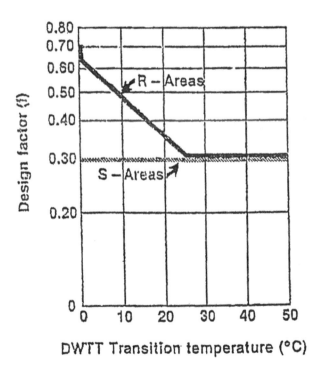

FIGURE 4.6 DWTT transition temperature (Source: Kadir, Ali. 2002. *Distribution, transmission system and design.* Lecture notes, Salford: School of Computing Science & Engineering – University of Salford).

for a minimum of 75% shear area when subject to drop weight tear test (DWTT), i.e. less than 25% brittle fracture. However, if the pipeline is to operate permanently at a level below 30% SMYS, then a DWTT is not required because propagating fractures do not occur below this level.

For existing pipelines operating above 30% SMYS, the maximum design factor should be that corresponding to the DWTT transition temperature, as shown in Figure 4.6. As previously described the arrest/propagate boundary is a function of the level of hoop stress and the transition temperature. Samples of the pipeline steel are required for DWTT before Figure 4.6 can be used.

Ductile fracture is avoided by ensuring that the pipeline steel has sufficient energy absorption properties to prevent fracture occurring. The Charpy V-notch impact test is used as previously described, and the energy level required depends on the grade of steel and the diameter of the pipe.

Drop weight tear testing is a material characterization test aimed at avoiding brittle fracture and ensuring crack arrest in pipelines (seamless or welded). Similar to fracture toughness testing, DWTT measures both the crack initiation and crack propagation characteristics of metallic materials at operating temperatures. The drop weight tear test (DWTT) is specified in API RP 5L3 or ASTM E436.

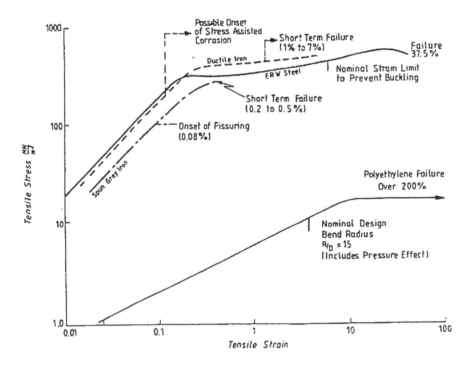

FIGURE 4.7 Typical stress–strain curves for various pipe materials (Source: Kadir, Ali. 2002. *Distribution, transmission system and design.* Lecture notes, Salford: School of Computing Science & Engineering – University of Salford).

There are four characteristics that can be used to assess the qualities of a material to serve as a pipeline:

- Strength
- Ductility
- Toughness
- Resistance to corrosion

Strength is usually shown diagrammatically in the form of a stress–strain curve for the material. Figure 4.7 shows the stress–strain characteristics of the common pipe materials.

Ductility is a measure of a material's resistance to bending stresses. Low-ductility materials, such as cast iron, break very easily under the influence of bending stress, whereas high-ductility materials, such as polyethylene (PE), are very resistant to bending, as shown in Figure 4.8.

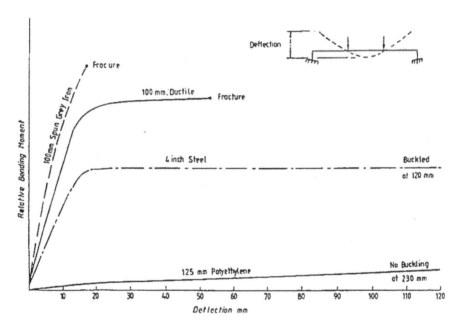

FIGURE 4.8 Bending characteristics for 2 m lengths of various pipe materials (Source: Kadir, Ali. 2002. *Distribution, transmission system and design.* Lecture notes, Salford: School of Computing Science & Engineering – University of Salford).

Toughness is a measure of a material's resistance to impact damage and its ability to resist fracture. Figure 4.9 shows test results for various resistances to impact, from different types of excavation tools (picks), for various pipe materials.

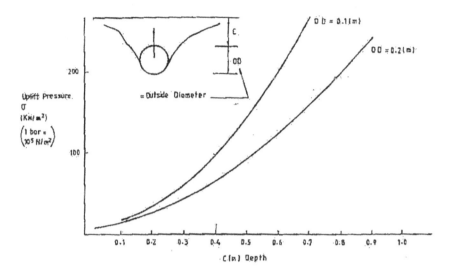

FIGURE 4.9 Impact energy to cause leakage in various pipe materials (Source: Kadir, Ali. 2002. *Distribution, transmission system and design.* Lecture notes, Salford: School of Computing Science & Engineering – University of Salford).

4.2.3.3   Material Selection

Selection of materials should not only look at the different types of steel available today but also the use of other materials such as plastics and fiber glass, lined steel pipe versus alloy steels, and the external coating material depending on the service requirements. This is covered elsewhere in the book. Each material has its own advantages that need to be considered in the design.

For instance, polyethylene (PE) pipe is now manufactured up to 16 bar rating and does not corrode in water; hence no coatings or cathodic protection is required, but the pressure rating reduces dramatically above 20°C and PE can leach hydrocarbon liquid out of or into the pipe.

4.2.3.3.1   Steel

Steel is well known all over the world; its various characteristics are well defined and standardized. However, it is advisable to pay attention to the quality of the steel and see to it that the various elements remain within prescribed limits, in relation to not only carbon, manganese, silicon, sulfur, and phosphorous but also aluminum, nitrogen, boron, chromium, copper, and molybdenum.

Steel has both strength and ductility (Figure 4.7), which means as a pipeline it can withstand very high internal pressures (>100 barg).

It is also resistant to bending (Figure 4.8) and impact (Figure 4.9). However, it does have weakness in that it is vulnerable to corrosion, and at very low temperatures it loses its toughness and ductility. These factors have played a major part in determining how steel can be used within pipeline systems.

The product of corrosion of steel does not adhere to the parent material but flakes away from it. Hence a steel pipe can fail to contain gas when the corrosion process penetrates the wall thickness of the pipe. This then allows gas to escape from the pipe initially at a low rate. On the other hand, general corrosion of a long length of a high-pressure pipeline can reduce the wall thickness to such a level that the parent material can no longer accommodate the stress in the pipe wall and then a rupture might take place. Also, there is a combination of high temperature (e.g. from gas compression) plus a corrosive environment which can give rise to a form of stress corrosion cracking.

The philosophy used in IGE/TD/1 is to minimize the exposure to a corrosive environment by using a protective coating, some form of active corrosion protection (i.e. sacrificial anode or impressed current) plus monitoring and maintenance of any cathodic protection system.

Since 1968, steel pipes have been coated with polyethylene either by extrusion or by powdering. This coating is seldom specified or subject to quality control test, whereas its expected life shall determine that of the steel pipe. The quality of the coating is conditioned by two performance parameters:

- Adherence to the steel pipe in order to avoid loosening and the formation of air-locks, which through their vapor permeability will involve corrosion, which will be hard to stop by cathodic protection, given the low current intensity proper to such coating
- The intrinsic quality of polyethylene as a material, its extrusion quality, and the amount of stabilizer and antioxidants

Conclusively, steel pipes manufactured according to the appropriate standards will suit for pressures of over 7 bar provided that:

- Attention is paid to the steel's quality.
- The pipes are passively protected by a polyethylene coating of good quality (equal in performance of PE pipes).
- The coating is accurately repaired with a compatible product at the place of welding and at any point of coating damage.
- There is cathodic protection of the pipe system and systematic supervision.

4.2.3.3.2 Cast Iron

Relatively speaking, cast iron is a weaker, and generally more brittle, material than steel (Figure 4.7). Accordingly, it is less able to withstand applied loads. Cast iron and steel corrode in a similar manner, but grey cast iron does have an advantage in that the corrosion products are contained in a graphite, carbide, and phosphide matrix that will contain gas at relatively low pressures. Unfortunately, the brittle characteristics of cast iron mean that, when iron pipes break, large quantities of gas can be released, even at low pressure.

There are four types of cast iron, commonly referred to as:

- Pit
- Sand spun
- Metal mold spun
- Ductile

The first three of these categories are classed as "grey iron" and the last as "spheroidal graphite iron". The only cast iron pipe material now commercially available in most countries is ductile iron; however, in Britain the use of all cast iron pipe materials ceased in 1974.

In Britain, grey cast iron pipe was laid over many years, and has, in general, served the industry well. Because of the wall thickness and the fact that the corrosion products were able to contain gas at low pressures, only a notional attempt was made to provide corrosion protection to the grey iron. Thus, it would be environmentally and economically unjustifiable to perform retrospective remedial measures. All that can be performed is the continual monitoring and maintenance of the system that IGE/TD/3 advocates. Unfortunately, the material has a low tolerance to the strain of any kind, as can be seen in Figures 4.6 and 4.7. The dominant mode of failure is to break in bending.

Two particular forms of cast iron pipe suffer from a stress-assisted form of corrosion. The worst of the two is metal molded, water-cooled, and centrifugally spun-cast iron where the defect produced by corrosion is a sharp notch or fissure. Almost all the failures of this material exhibit fissure corrosion on the fracture faces. Again, the most common failure mode is a circumferential fracture in bending.

Spheroidal graphite (commonly known as ductile) iron pipe exhibits a similar form of stress corrosion, but the shape of the notch is more rounded. A further point which must be considered is that the products of corrosion are less integrated with the spheroidal graphite matrix and do not create a material which will readily contain the internal gas pressure. The most common failure mode is the perforation of the pipe wall by corrosion rather than the fracture type of failure associated with other forms of cast iron.

For new systems, IGE/TD/3 proposes that consideration and active measures be taken to minimize the effects of the environment as well as the loads to which any pipe is subjected.

4.2.3.3.3   Plastic Materials

Plastics are fairly new materials whose quality is still being developed. Plastics have been used to carry gas since 1950. They were first used in the United States where various studies were carried out, resulting in networks of various types of plastic, e.g. ABS, CAB, PVC, CPUC, PA, and PE.

In Europe, the Netherlands was the first country to use plastic, opting for PVC in 1954, while Belgium and the United Kingdom started using PE in 1968. British Gas later turned to American materials and technology, while Belgium preferred from the outset to develop its own system.

Gaz de France has been using polyethylene for connections since 1974 and for the construction of primary networks since 1978. Gaz de France uses a medium pressure 4 bar distribution system with electrofusion fittings.

4.2.3.3.4   Polyethylene

Polyethylene (PE) pipe is the weakest in terms of the ultimate tensile strength of the materials to be considered here, as shown in Figure 4.7, but it is the most flexible material and has a considerable elongation at failure. It is not vulnerable to galvanic corrosion, but it has time-dependent strength properties. These time-dependent properties will eventually lead to a failure to contain gas, but experience to date from the United States and Britain indicates that its life is well in excess of the predicted design minimum of 50 years.

Improvements in PE pipe technology have produced materials that are capable of working at 10 barg with good safety factors built in and future developments, such as PE-XL will extend this range of working pressure. Because it does not corrode, it can be easily joined by automatic fusion systems and has many other operational advantages. PE is the first choice of material for most gas companies. However, the fact that PE 80 can operate up to 5 barg and PE 100 can operate up to 10 barg does not mean that these pressures are acceptable in every location at that pressure. Safety and risk issues must also be considered (Table 4.2).

TABLE 4.2
PE Classification

	Previous classification	Current ISO classification
First generation	PE 50	MRS 63
Second generation	PE 63	MRS 80
Third generation	PE 80	MRS 100

4.2.3.3.5   Fittings

Whichever material is selected must have fittings to provide flexibility in the geometry of design and linking of different materials. These fittings have a range of methods of connection, and it is this characteristic which is most important in assessing performance. Ideally, a fitting should not reduce the strength or integrity of the system; that is, it should not be the weak link.

4.2.3.3.5.1   Welded Fittings   Within this category, the joining of fittings to both steel and plastic will be considered. Both require the raising of temperatures to enable the fusing together of similar material, and both methods, if successful, produce a joint which is as strong as, if not stronger than, the original material. This applies only if the operation is done properly. Steel electrodes and welding procedures plus non-destructive testing (NDT) techniques mean that all welds can be checked and errors can be determined relatively simply.

Unfortunately, this is not the case with plastic pipe. Procedures and codes of practice are available, but the quality assurance check is limited to visual inspection of the joint. At the moment, no fully proven NDT technique has been found, although there are some reasonable radiography and ultrasonic methods. There are three types of joints for PE: socket fusion, butt fusion, and electrofusion. All three joints provide strength equivalent to the pipe if loaded without bending. However, the socket-fused joint can fail under bending and should not be used. It is important that the jointing equipment be properly maintained in order to achieve the best results. The introduction of electrofusion fittings has helped remove the majority of operator errors and tool malfunction.

Failure of these fittings will have similar characteristics to the material, i.e. a ductile tear in the case of steel and crack growth in the case of PE (Kadir 2002).

4.2.3.3.5.2 Threaded Fittings This system of joining pipes can only be performed with materials that are tolerant of being tapped; that is, they are tolerant of points of stress concentration. This essentially means that they are tolerant of points of stress concentration and that they are limited to metals. A screwed fitting provides a good mechanical joint in terms of strength but has three weaknesses: leakage through badly formed threads, the possibility of inferior materials joining two "higher-quality" pipes, and enhanced corrosion at the thread roots if not protected.

The failure modes of threaded joints are:

- Leakage through poorly made thread forms
- Leakage through drying out of thread filling material
- Leakage through corrosion holes in the root of the threads if not protected
- Fracture of inferior material or threads in the pipe or coupling pieces

The first three forms release only small quantities of gas whereas the fourth can release large quantities. However, threaded fittings are rarely used above 50 mm (2 in) diameter.

4.2.3.3.5.3 Bolted Fittings There are several proprietary jointing arrangements using this principle, but all of them depend upon some sort of gasket that is compressed to achieve a gas-tight seal. Depending upon the design, this joint is mechanically strong. Unfortunately, the gasket materials relax with time, and a small leakage of gas can occur. Some materials, like natural rubber, are adversely affected by the dry gas that is transmitted. This can result in the shrinkage of the gasket, again resulting in small quantities of gas escaping.

In addition, the failure of bolts through corrosion or the pull-out of non-end load-bearing joints can give rise to the release of large quantities of gas. Despite these facts bolted fittings have been used successfully for many years in many gas companies.

4.2.3.3.5.4 Compression Fittings These joints rely on the compression of either the pipe wall or a component like a copper olive to provide a grip from one pipe through the coupling to the next pipe. Depending on the design, these joints can tolerate considerable load and deflections but tend to fail by the compression component being pulled over the pipe. This is a rare occurrence for normal operations, but mechanical interference can result in fittings being pulled off. Compression fittings for PE pipe are designed to be stronger than the pipe and hence should not fail.

4.2.3.3.5.5 Glued Fittings Glued joints are used on nylon fittings for gas distribution pipe. However, there is a health and safety problem for operatives exposed to solvent-based glues, and it is very difficult to control or check the quality of the joint;

therefore this type of joint is not recommended. The failure of the jointing compound generally releases only small quantities of gas.

4.2.3.3.5.6   Comments on Joints   A system is only as "strong" as its weakest component. There is an inherent weakness in any system that incorporates joints. This weakness is that the joint is made by humans, and its reliability must therefore be questioned.

4.3   QUALITY CONTROL OF PIPELINE CONSTRUCTION

It is often said that doctors bury their mistakes; the same can easily be said of pipe-liners. The only way to reduce material and construction defects is with effective quality control. Material quality is greatly improved over the condition from 10 to 15 years ago, partly due to automatic inspection techniques and the application of standards such as ISO 9000. The use of independent inspectors for both materials and construction checks such as a welding and field coating is becoming less common and allows the possibility of an increasing amount of defects to be undetected or the margin to be drawn in favor of suppliers or contractors.

The main method of checking welding by NDT is the use of X-ray or gamma-ray shot through the pipe from one side to another. Normally at least two films will be required to get full coverage, but depending on the size of the pipe this could be three or more. The use of an internal source is possible but relatively rare. The pipe needs to be empty to avoid blurring of the films, and field joint coating normally awaits the outcome of an X-ray to avoid rework in the event of a repair.

Weld procedures need to be written stating the parameters including the type and size of the rod, pre-heat required, and other variables. This forms the weld procedure. Each weld procedure is then normally destructively tested to prove that the procedure results in a weld of suitable strength. The results of such a test are written on a Weld Procedure Qualification Report (WPQR). The welder(s) who performed the test are then qualified to perform that procedure. Other welders who perform the procedure that has been tested need only do a shortened test, sometimes the first production weld, to be qualified, providing no repairs result. Control over the welding consumables, particularly with respect to baking them and heating them prior to use, is very important in keeping weld repairs to a minimum (Kadir 2002).

4.3.1   PIPE JOINTING

As with materials, jointing design in design codes and elsewhere concentrates on the welding of carbon steel pipelines, which is by far the most common form of jointing seen on pipelines worldwide. However, alternative techniques are available for non-welded connectors, including push-fit connectors and screwed connectors for steel pipelines and various fusion welded and epoxy glue connectors for PE and fiber-glass pipelines (Kadir 2002).

Metallic screwed and push-fit connectors can deliver fast lay rates of more than 1 km per day for a single team, regardless of wall thickness, and can use high strength and difficult-to-weld steels including line pipes. Due to the additional cost of the connector and other costs that are still incurred regardless (clearing, ditching, back-filling, etc.), it is unlikely that these connectors will ever be cheaper than welding normal carbon steel pipe. However, some of the technical advantages of non-welded connectors can be used to swing the advantage to the use of such connectors and are worth considering in the design of some pipelines.

These can be summarized below:

- Jointing rate independent of wall thickness
- Joining internally lined or coated pipe without post-connection work or removal of the coating in the weld area
- Use of very high strength (X80+) or high chrome (13%) pipe without regard to welding difficulties
- No X-ray requirement
- Making use of cheaper, semi-skilled labor.

Their disadvantages are:

- Additional cost of connectors often makes their use cost neutral or more expensive.
- Connectors are not yet recognized as an industry standard for pipelines.
- Legislation or design code may state welding is required, and hence a dispensation needs to be obtained.
- Clients and engineers may have a conservative reaction.
- There may be a limited number of "patented" suppliers.

4.3.2 MITIGATION FROM IMPROPER DESIGN OR MATERIALS SELECTION DURING OPERATION

If failures, direct examinations, and/or risk assessments identify a high susceptibility to unacceptable design or materials, one or more of the following types of risk mitigation actions shall be initiated:

- In-line inspection of the pipeline to determine the nature and extent of design and material issues
- Pressure test of the pipeline to identify weld or material flaws

5 Geotechnical Hazards

5.1 MITIGATION OF GEOTECHNICAL HAZARDS

Geological engineers use the term "Geohazard" to describe the hazards to the pipeline that may derive from any potential gravity-related geological/geotechnical problem or failure. Geotechnical hazards threaten pipelines across the world. Modeling, identifying, mitigating, and monitoring hazard-prone areas can save millions of dollars in repair, replacement, and lost service. Common Geohazard that affect pipelines include:

- Landslides/slope instability
- Ground settlement
- Erosion
- Fault movement

Onshore, the most important are volcanic eruptions, earthquakes, landslides and debris flows, floods and snow avalanches. Offshore, slope instability and earthquakes are the main threats because of their potential for damaging seafloor installations, and for generating devastating tsunamis, such as the 1998 Papua New Guinea event responsible for more than 2000 deaths, and the past Storegga Slide tsunami.

It is evident that the safety of any pipeline is directly related to

a) the verification of the pipeline against the identified Geohazard, and
b) the proposal and the design of any mitigation or protection measure in case of excessive pipeline distress.

Key ingredients of the mitigation of pipeline Geohazard include:

- Geohazard assessment: Geohazard identification and characterization
- Re-routing
- Pipe–soil interaction analysis
- Strain-based design criteria
- Weld strength and quality
- Relate pipeline design and construction standards to the degree of ground shaking risk.
- Adopt ordinances that require geologic and seismic site investigations before development proposals can be approved.

Regarding seismic loads, examination of papers and design guidelines shows that for a welded steel pipeline, ground shaking and strain do not cause any substantive harm. Problems for buried pipelines arise either where the pipeline crosses a fault with a surface expression or where the ground in which the pipeline is laid loses strength and a *landslide* or *slip* leaves the pipe without support or imposes large shear forces when moving. The effect of fault movement can be mitigated by design-ing long straight lengths of pipes crossing the fault at an angle of 60 degrees and backfilling the pipeline trench with a cohesionless material (e.g. gravel). The follow-ing mitigation strategies are suggested.

5.1.1 Geotechnical Investigation

The geotechnical investigation provides sub-surface data along the pipeline route, including the soil and rock engineering data that is required for the pipeline design and construction (IPLOCA 2014). A geotechnical investigation comprises site and laboratory operations that may include:

a) Exploratory holes (in addition to detailed description (logging) of samples, drill cores, and exposed soil and rock faces)
   • Boreholes
   • Cone penetration tests
   • Trial pits
   • Trial trenches
b) Geophysical investigations
   • Surface investigations using (for example) shallow seismic and resistiv-ity methods
   • Borehole investigations using geophysical tools in boreholes
c) In situ geotechnical testing
   • In boreholes, for example standard penetration tests, pressuremeter tests, permeability and pumping tests
   • With cone penetration tests, such as piezo-cone
   • In Trial pits and trenches, for example density tests and scan-line sur-veys of rock fractures
d) Geotechnical monitoring
   • Groundwater conditions, e.g. piezometers in boreholes
   • Ground movements, e.g. inclinometers in boreholes, surface movement markers
e) Laboratory testing (of samples and drill cores of the ground and of ground water)
   • Index tests, such as grain size, plasticity, moisture content, mineral composition
   • Shear strength of soils, tested in drained and undrained conditions, including tests to determine residual shear strength
   • Compressive strength of rocks, including the point load index test and tests of residual shear strength
   • Compressibility
   • Compaction, including the use of soil and rock as fill materials
   • Durability of rocks for use as fill, aggregate and amour stone
   • Permeability
   • Geochemistry and groundwater chemistry, including aggressive and contaminated ground
   • Thermal and resistivity properties

5.1.2 GEOHAZARD ASSESSMENT

Identification of potential *landslide* areas will need to form part of the route survey in seismically active areas, and susceptible areas should be avoided wherever possible (IPLOCA 2014).

Geohazard assessment provides information on earth surface processes and geological hazards that might pose a threat to pipelines and associated infrastructure. Geohazard assessment of a project site should cover the following:

a) Desk study

Perform a desk study to describe the geologic setting along the alignment and conduct a preliminary assessment of the Geohazard. The main inputs to the desktop study include aerial photography, geologic maps from government agencies and local authorities, and data from open sources (i.e. province boundaries, district boundaries, roads, and rivers).

b) Mapped earthquake faults and seismic zones
   - Earthquake vibrations
   - Ground liquefaction, including lateral spreading
   - Fault rupture
   - Tsunami

c) The intensity of ground shaking at the project site determined by probabilistic methods (10% probability of occurrence in 50 years)
   - Ground settlement and collapse:
     - swelling, shrinking ground and collapsible soils,
     - ground solution, and collapse (karst)

d) Potential for liquefaction, ground failure, and landslides at the site
   - Landslides and erosion: Pre-existing landslides and erosion areas and landslide-prone and erosion-prone terrain

e) Potential for flooding at the site from man-made facilities and natural storms
   - Coastal erosion and deposition
   - River behavior, including channel changes and bed erosion
   - High water table and flooding

f) Contaminated land and ground geochemistry of the project site, including methods for mitigation:
   - Naturally aggressive ground and groundwater
   - Former and current industrial uses
   - Contamination of the soil and rock
   - Contamination of groundwater and surface water

5.1.3 RE-ROUTING

Pipelines have been built in seismically active areas as both buried and surface-run pipelines. In general, there is little advantage to having the pipeline above ground as general shaking is not a particular problem for buried pipelines, but pipelines have been known to fall off supports or be excessively strained at individual supports, producing a point load, which is not associated with buried pipelines.

TABLE 5.1

Pipeline Geohazard (IPLOCA 2014)

Geo-hazard	Description	Routing mitigation
Landslide	Ground displacement and movement of a mass of rock, earth or debris down a slope	• Avoid if possible • Minimize sidelong routing across the landslide, route parallel along the axis of ground movement
Gullying, soil erosion and fluvial erosion	Removal of soils by water, wind or ice action or by down-slope scree	• Avoid areas of active erosion if possible • Minimize sidelong routing parallel to erosion area, cross at 90°
Mobile sand dunes	Fragile desert habitat that maybe damaged or blown away by wind	• Avoid if possible • Minimize crossing length
Earthquakes and fault lines	A fracture in the continuity of a rock formation caused by a shifting or dislodging of the earth's crust, in which adjacent surfaces are displaced relative to one another and parallel to the plane of fracture.	• Avoid if possible • Special design considerations (eg. Finite element analysis) will be required if un-avoidable • Special/engineered backfill techniques likely to prevent pipe damage during an earthquake (such designs are common in areas like Japan)
Volcanoes	The vent and the conical mountain left by the overflow of erupted lava, rock	• Avoid • Avoid existing flow canals
Soft soils	Soils that may not be able to support a pipeline (swamp, peat, bog)	• Methods to cross soft soils include support anchors screwed into hard soil below the soft soil; support mattresses under the pipeline to reduce bearing pressure; neutral buoyancy to ensure that pipe neither sinks or floats after installation. It may also be possible to remove weak soil and replace with engineered backfill.
Underground cavities	Areas of coal mining, caves, caverns, subsidence areas	• Avoid if possible • Methods to design for and cross underground cavities are possible. These include pumping concrete into the underground mines (subject to size and volume), the whole mine need not be filled in, but sufficient to limit settlement; use thicker wall pipe acceptable for estimated settlements
River channel migration	River banks erosion leading to river meander, and river bed erosion leading to bed channels of varying depth	• Feasible to estimate and design for river meander and river bed erosion. This will generally include sufficient burial in river bed, and sufficient deeper burial extent from river banks. • River bank erosion prevention methods can also be used. • Minimize crossing length
Aggressive soils	Contaminated soils	• Avoidance will depend on type of contamination, and if disturbed the safety impact on local population and works: environmental impact and disposal issues. • Minimize crossing length

In the event of mass ground failure, both types of construction are equally vulnerable and the pipeline would be fortunate to survive. Where other considerations apply, supports have been specifically designed to move only under large forces which would be close to the yield strength of the pipeline. Project examples are rare as particular fault areas are normally avoided wherever possible, and where ground failure is predicted from soil analysis, re-routing is normally undertaken.

Although rarely possible, Geohazard should be avoided by a pipeline route as far as possible. Table 5.1 describes various routing mitigating measures proposed by the International Pipe Line and Offshore Contractors Association (IPLOCA) in 2014.

5.1.4 STRAIN-BASED DESIGN

Strain-based design is appropriate where the stresses and strains exceed the proportional limit and where the peak design loads will be reduced when the material strains.

Strain-based design has proven applicability to pipelines laid offshore, pipelines operating at high temperatures, pipelines in the areas of soil movement, and arctic pipelines.

Soil movement should generally be considered displacement controlled. However, situations exist where soil-induced loadings are load controlled or intermediate between load and displacement controlled (Douglas et al., 2008).

It is generally not practical to design pipeline crossings of permanent ground displacement (PGD) zones (i.e. high-strain conditions) in accordance with the usual code-allowable stresses for operating and external load conditions. Instead, the accepted alternative approach is to revert to a strain-based design that allows comparatively large strains to occur locally in the pipeline, provided that the pressure boundary integrity is assured, i.e. allows damage to the pipe but prevents pipe rupture and the release of contents. Tensile strain limits may be taken as high as 3%–4%, depending on the quality of the girth welds. Compressive strain limits typically range from about 2% to 4% for the range of diameter-to-thickness ratio (D/t) of 90:45. Strain acceptance limits are based on the assumption that the pipeline girth welds will be capable of developing gross section yielding of the pipe wall. This capability, often referred to as "overmatching welds", means that failure would occur in the pipe before failure occurs in the weld or the weld heat-affected zone. This implies that a welding process and weld inspection program to minimize both the number and size of imperfections would be implemented during construction. The principles of strain-based design are well chronicled in the technical literature and will not be repeated here. Likewise, the methodology for performing the non-linear finite element analysis of pipelines subjected to PGD is well established and described in available industry guidance documents.

For active faults and liquefaction zones, a pipeline must be designed to withstand the full effect of the PGD without rupture, as there is no time for intervention once the event is initiated (Douglas et al., 2008).

For landslides, knowing the limitations of strain-based design for landslide hazard mitigation, it is essential that a monitoring program be established to detect ground movements prior to reaching damaging levels, which might vary among respective sites. Ideally, the extent and scope of monitoring should be based on site-specific risk assessments of the inventory of landslide areas, subject to adjustments based on the findings and conclusions from periodic observations and evaluation of data (Douglas et al., 2008).

5.1.4.1 Determination of Component of Strain

Strain-based design is a design method that places a limit on the strains at the design condition rather than the stresses.

The structural analysis of pipeline systems is concerned with the determination of the stress or strain state of a pipeline and the subsequent check of the stress or strain state against a failure criterion. The general term of structural analysis is indeed very wide, and this chapter does not attempt to offer a complete treatment of structural analysis but considers in some detail the more important aspect in relation to pipeline systems.

The chapter aims to provide guidance, should a more accurate treatment of strain state in a pipeline be required. Traditionally it is acceptable to make certain conservative assumptions to simplify the structural design of pipelines, especially when there is limited data available for the problem. In some situations, the conservatisms adopted may require further consideration to refine the solution. It is important in this case that the designer is aware of the fundamentals of mechanics of the materials and some simple structural concepts. It should be very clear that the strain state of a pipeline can be given only for a specific point or plane in the material. Therefore, to be consistent when the individual strain components are checked against a failure criterion, the following must be true:

- Each stress or strain component should have the maximum (or principle) value. This does not always coincide with the traditional x, y, and z axes (longitudinal, hoop, and radial directions, respectively).
- These stresses or strains should apply to the same point in the material (for example, the inner surface of pipe). This can be very important, especially in a three-dimensional strain state, where one component's strain has a maximum of the inner surface and another perpendicular strain has a maximum on the outer surface.

The emphasis for the structural design of a pipeline is on a stress-based analysis. A strain-based approach is limited to areas where a simple strain argument provides a useful additional tool to the engineer.

The primary areas where the strain-based design will be used are in the design of reeled laying of offshore pipelines, in the thermal design of arctic pipelines, in the design of offshore pipelay systems, in the design and assessment of pipelines in areas with significant expected ground movement, and in high-temperature and high-pressure (HT/HP) pipeline designs.

A pipeline may also have some applications of strain-based design where cyclic loadings cause occasional peak stresses above the pipe yield strength. Here, the cyclic lifetime assessment is improved by using strain ranges for the cycles, instead of stress ranges.

During operation, the limit on equivalent stress may be replaced by a limit on allowable strain, provided that all the following conditions are met.

The limit on equivalent stress recommended in BS 8010, Part 2, Clause 6.4.2.4 may be replaced by a limit on allowable strain, provided that all the following conditions are met.

a. The allowable hoop stress criterion (see BS 8010, Part 2, Clause 6.4.2.1 and 6.4.2.2) is met.
b. Under the maximum operating temperature and pressure, the plastic component of the equivalent strain does not exceed 0.001 (0.1%). The reference state for zero strain is the as-built state (after pressure test). The plastic component of the equivalent uniaxial tensile strain should be calculated as:

$$\varepsilon_p = \left[\frac{2}{3} \left(\varepsilon_{pL}^2 + \varepsilon_{ph}^2 + \varepsilon_{pr}^2 \right) \right]^{1/2} \tag{5.1}$$

where
ε_p = equivalent plastic strain
ε_{pL} = principal longitudinal plastic strain
ε_{ph} = principal circumferential (hoop) strain
ε_{pr} = radial plastic strain

This analysis can be performed conservatively by assuming a linearly elastic–perfectly plastic stress/strain curve. Other, more realistic stress/strain curves may be used. However, it is essential that the assumed curve is validated as being conservative by material stress/strain curves from the manufactured pipe.

c. Any plastic deformation occurs only when the pipeline is first raised to its maximum operating pressure and temperature but not during subsequent cycles of depressurization, reduction in temperature to the minimum operating temperature, or return to the maximum operating pressure and temperature. This should be determined via analytical methods or an appropriate finite element analysis. The analysis should include an estimate of the operational cycles that the pipeline is likely to experience during the operational lifetime.
d. The D/t_{nom} ratio does not exceed 60.
e. Welds have adequate fracture resistance to accept plastic deformation when determined either by direct testing or by fracture mechanics testing and analysis. Where direct testing is employed, pipes containing maximum credible flaws located in the weld metal and heat-affected zone (HAZ) should show that fabrication flaws do not extend beyond acceptable limits when subjected to maximum operational loads. If design envisages cyclic loading, this needs to be anticipated in the test. Where fracture mechanics testing analysis is employed, testing of representative pipe welds should be conducted in accordance with BS 7448-1, BS 7448-2, BS 7448-4, and BS EN ISO 12737 as appropriate. The analysis procedures for fatigue and

fracture should be in accordance with BS 7910 in order to ascertain whether maximum credible flaws in the weld metal and HAZ extend beyond acceptable limits.

f. Actual or angular misalignment at welds is maintained within defined tolerances.

g. A fatigue analysis is carried out in accordance with BS 8010, Part 2, Clause 6.4.6.4.

h. A fracture analysis is carried out in accordance with Clause 6.4.6.4.–I of BS 8010, Part 2). Additional limit states are analyzed as follows:

1. Bending failure resulting from the application of a moment in excess of the moment capacity of the pipe
2. Ovalization – distortion of the pipe wall associated with bending to high-strain levels (see Clause 6.4.4.2 and Annex G)
3. Local buckling (see Clause 6.4.4.1 and Annex G)
4. Global buckling – lateral or upheaval buckling due to overall axial compression (see Clause 6.4.4.1 and Annex G)

Plastic deformation reduces pipeline flexural rigidity; this effect can reduce resistance to upheaval buckling and should be checked if upheaval buckling might occur.

The effects of strain localization should be taken into account in the strain-based design. Strain localization is associated with discontinuities in the stiffness of the pipeline (bending or axial) and can therefore develop in the following locations:

- Locations where wall thickness is changed
- Buckle arrestor locations
- Locally thinned regions, e.g. due to corrosion
- Field joints and coatings
- Welds, due to under-matching of the strength of the weld

5.1.4.2 Buckling

The following calculations are as per the requirements of BS8010 Part 2, Annex G.

5.1.4.2.1 Local Buckling

Local buckling of the pipe wall can be avoided if the various loads to which the pipe is subjected to are less than the characteristic values in Equations (5.2)–(5.20). Where the concrete cladding is thick enough and reinforced to provide a structural member conforming to BS 6349-1 and BS 8110, it may be used to provide support against buckling, provided that appropriate documentation is given.

5.1.4.2.1.1 External Pressure The characteristic value P_C that causes collapse when the external pressure is acting alone can be calculated using Equations (5.2)–(5.5).

$$\left\{\left(\frac{P}{P_e}\right)-1\right\}\left\{\left(\frac{P_c}{P_y}\right)^2-1\right\}=\frac{P_c}{P_y}\left(f_0\frac{D_0}{t_{nom}}\right) \tag{5.2}$$

$$P_e = \frac{2E}{\left(1-v^2\right)}\left(\frac{t_{nom}}{D_0}\right)^3 \tag{5.3}$$

$$P_y = 2\sigma_y \frac{t_{nom}}{D_0} \tag{5.4}$$

$$f_0 = \frac{D_{max}-D_{min}}{D_0} \tag{5.5}$$

where

P = external overpressure (buckling)
P_e = critical pressure for an elastic circular tube
P_c = characteristic external pressure (collapse)
P_y = yield pressure
f_0 = initial ovalization of a pipe cross-section
D_0 = outside diameter of a pipe
t_{nom} = nominal wall thickness
E = Young's modulus of elasticity
v = Poisson's ratio
σ_y = SMYS of the pipe wall material
D_{max} = maximum (oval) outside diameter
D_{mim} = minimum (oval) outside diameter

5.1.3.2.1.2 Axial Compression If D/t_{nom} is less than 60, local buckling under axial compression does not occur until the mean axial compression load F_{xc} reaches the yield load F_y, i.e. as shown in Equation (5.6).

$$F_{xc} = F_y = \pi\left(D_0 - t_{nom}\right)t_{nom}\sigma_y \tag{5.6}$$

where

F_{xc} = mean axial compressive load
F_y = yield load

All other parameters are as defined above.

5.1.3.2.1.3 Bending The characteristic bending moment value M_c required to cause buckling when bending moments are acting alone can be obtained using Equations (5.7) and (5.8).

$$\frac{M_c}{M_p} = 1 - 0.0024\frac{D_0}{t_{nom}} \tag{5.7}$$

$$M_p = \left(D_0 - t_{nom}\right)^2 t_{nom}\sigma_y \tag{5.8}$$

The characteristic bending strain ε_{bc} at which buckling due to bending moments acting alone occurs can be obtained using Equation (5.9).

$$\varepsilon_{bc} = 12\left(\frac{t_{nom}}{D_0}\right)^2 \tag{5.9}$$

where

M_c = characteristic bending moment
M_p = full plastic moment capacity of pipeline cross-section
ε_{bc} = characteristic bending strain

All other parameters are as defined above.

5.1.3.2.1.4 Torsion The characteristic value τ_c that causes buckling when torsion is acting alone can be obtained using Equations (5.10)–(5.14). Equation (5.10) is used when $a_\tau < 1.5$, Equation (5.11) is used when a_τ is ≥ 1.5 and ≤ 9, and Equation (5.12) is used when $a_\tau > 9$.

$$\tau_c / \tau_y = 0.542 x a_\tau \tag{5.10}$$

$$\tau_c / \tau_y = 0.813 + 0.068\left(a_\tau - 1.5\right)^{0.5} \tag{5.11}$$

$$\tau_c / \tau_y = 1 \tag{5.12}$$

$$\tau_y = \frac{\sigma_y}{3^{0.5}} \tag{5.13}$$

$$a_\tau = \frac{E}{\tau_y}\left(\frac{t_{nom}}{D_0}\right)^{3/2} \tag{5.14}$$

where

τ_c = characteristic torsional shear stress
τ_y = yield shear stress
a_τ = torsion coefficient

All other parameters are as defined above.

5.1.4.2.1.5 Load Combinations The maximum external overpressure, P, in the presence of compressive axial force, F_x, and/or bending moment, M, when f_0 is less than 0.05 (5%) can be calculated using Equation (5.15), where:

- γ is calculated using Equation (5.16).
- σ_{hb} is calculated using Equation (5.17).
- σ_{hcr} is calculated using Equation (5.18) or (5.19) as appropriate.

$$\left\{\left(M/M_c\right) + \left(F_x/F_{xc}\right)\right\}^\gamma + \left(P/P_c\right) = 1 \tag{5.15}$$

$$\gamma = 1 + 300 \times \frac{t_{nom}}{D_0} \times \frac{\sigma_{hb}}{\sigma_{hcr}} \tag{5.16}$$

$$\sigma_{hb} = \frac{P \times D_0}{2t_{nom}} \tag{5.17}$$

$$\sigma_{hcr} = \sigma_{hE} = E\left(\frac{t_{nom}}{D_0 - t_{nom}}\right)^2 \quad \text{for} \quad \sigma_{hE} \leq \frac{2}{3}\sigma_y \tag{5.18}$$

$$\sigma_{hcr} = \sigma_y\left\{1 - \left(\frac{1}{3}\right) \times \left(\frac{2\sigma_y}{3\sigma_{hE}}\right)^2\right\} \quad \text{for} \quad \sigma_{hE} > \frac{2}{3}\sigma_y \tag{5.19}$$

Values for P_C, F_{xc}, and M_c can be obtained from Equations (5.2), (5.6), (5.7), and (5.8), respectively.

Where

γ = factor used in the calculation of load combinations

σ_{hb} = hoop stress used in buckling analysis

σ_{hcr} = critical compressive hoop stress when pressure is acting alone

σ_{hE} = critical compressive hoop stress for completely elastic buckling

All other parameters are as defined above.

5.1.4.2.1.6 Strain Criteria The bending strain, ε_b, required to cause buckling in the presence of external overpressure, P, can be calculated using Equation (5.20).

$$\frac{\varepsilon_b}{\varepsilon_{bc}} + \frac{P}{P_c} = 1 \tag{5.20}$$

Values for ε_{bc} and P_c can be obtained from Equations (5.9) and (5.2), respectively.

Where

ε_b = maximum bending strain

ε_{bc} = characteristic bending strain

All other parameters are as defined above.

5.1.4.2.2 Propagation Buckling

The potential for a pipeline to propagate local buckles is dependent on the external overpressure, P, and its relationship with the propagation pressure P_p. The external overpressure, P, can be calculated using Equation (5.21).

$$P = P_0 - P_i \tag{5.21}$$

The propagation pressure P_p can be calculated using Equation (5.22).

$$P = 10.7\sigma_y\left(t_{nom} / D_0\right)^{2.25} \tag{5.22}$$

where

P_0 = external pressure

P_i = internal pressure

If P is less than P_p, then, even though it is possible for the pipe to develop a local buckle, the buckle will not propagate. If P is greater than or equal to P_p and a local buckle or local damage has occurred, then the pipeline is likely to undergo propagation buckling. It can be advisable to provide buckle arresters at strategic locations along the pipeline to limit the amount of pipeline damaged by a propagated buckle.

5.1.4.2.3   Upheaval Buckling

Two major factors contribute toward the upheaval of subsea pipelines: the effective axial driving force, arising from the internal pressure and temperature, and the presence of vertical out-of-straightness (OOS) in the seabed profile. The resistance to upheaval is provided by the submerged weight of the pipeline, plus any overburden, if present. The effects of the various parameters may be gauged by considering the equilibrium of the system at the point of instability. It is instructive to examine the upheaval phenomena in terms of the dimensionless coefficients. A typical imperfection is defined in terms of a characteristic length, L_C, and an imperfection height, H_e. The download coefficient, φ_w, is calculated using Equation (5.23).

$$\varphi_w = \frac{wE \times I}{H_c P_{eff}^{2}} \tag{5.23}$$

Similarly, the imperfection length may be characterized by a dimensionless coefficient using Equation (5.24).

$$\varphi_L = L_c \left(\frac{P_{eff}}{E \times I} \right)^{0.5} \tag{5.24}$$

It can be shown, from either numerical or experimental test results, that there exists a functional relationship of the form given in Equation (5.25).

$$f\left(\varphi_w, \varphi_L\right) = 0 \tag{5.25}$$

The functional relationship shown in Equation (5.25) may be used to assess the overburden requirement to prevent the upheaval buckling of the pipeline.

The following issues need to be taken into account when determining the overburden requirement:

a) Spacing of the vertical profile data
b) Uplift resistance of the backfill, remolded clay, backfilled sand, etc.

The stress/strain level along the pipeline should be within allowable limits, and remedial work should be carried out where these limits are exceeded.

Where

w = submerged weight of pipe plus overburden
P_{eff} = effective axial force

I = second moment of area
H_c = characteristic upheaval buckle imperfection height
φ_L = imperfection coefficient

All other parameters are as defined above.

5.1.4.2.4 *Ovalization*

The total ovalization, f, of a pipe due to the combined effects of unidirectional bending and external pressure can be calculated using Equations (5.26)–(5.28).

$$f = C_p \left\{ C_f \left(\frac{D_0}{t_{nom}} \varepsilon_b \right)^2 + f_0 \right\} \tag{5.26}$$

$$C_p = 1 / \left(1 - \frac{P}{P_e} \right) \tag{5.27}$$

$$C_f = 0.12 \left\{ 1 + D_0 / (120 t_{nom}) \right\} \tag{5.28}$$

Values for P_e and f_0 can be obtained from Equations (5.3) and (5.5), respectively.
Where:
C_p = ovalization magnification factor accounting for pressure effects
C_f = flattening coefficient

All other parameters are defined above.

NOTE: If cyclic or reversed bending is applied, the resulting ovalization can be considerably greater than that predicted by the equation.

5.1.5 SCHEDULING

Pipeline projects should schedule geologic investigations early in the preliminary engineering (FEED) stage to mitigate pipeline routing through geohazard areas.

Delay attributed to schedule and budget overruns in pipeline construction projects is a major challenge in the oil and gas industry. Some other factors include: poor managerial skill, slow decision making within all project teams; lack of communication between the client, consultant, and contractor; inadequate design team; scope variations; unrealistic contract decision; and delay in the preparation of the project drawings.

It is very important to create a clear project schedule. A project schedule results from knowing what the project will deliver; it should express all the requirements and the way they will be achieved in the form of tasks and activities. After the project schedule preparation, a contingency plan should be addressed to avoid any future delays.

Considering that the government is the owner of projects concerning the construction of pipelines in the oil and gas industry, it is vital that it integrates planning, scheduling, and budgeting training for nationals in the oil and gas sector. In addition, there is a need for the incorporation of senior expertise from other oil and gas countries to help in knowledge transfer and transfer of management and administration skills (Reyadh, Saad 2019).

There is a need for consistent monitoring and mitigation of contractor and owner-related factors, especially those with high-risk impact and frequency, such as inadequate project planning, budgeting, and scheduling; scope variations; and late materials delivery. This requires the development of project data about the root triggers of the critical delay factors.

5.2 MITIGATION FROM EXTERNAL FORCES

If routine ROW patrolling identifies any risks associated with ground movement, soil erosion, or river/creek bottom scouring, the following types of risk mitigation activities shall be initiated (Okyere 2015):

- Depth of cover and elevation survey of the affected section of the pipeline
- Assessing underground movement using monitoring equipment, such as inclinometers or strain gauges
- Hydrotechnical and/or geotechnical engineering evaluations to determine remedial action options, which may include:
  - Pipeline re-routing or replacement using horizontal directional drilling
  - Line lowering within the existing right-of-way
  - Armoring of approach slopes and banks to mitigate further damage

6 Off-Spec Natural Gas

6.1 INTRODUCTION

Off-spec gas means natural gas that does not conform to the specification (e.g. Water, sulfur, or condensate in a gas stream, products in a pipeline that are mixed). Hydrogen sulfide is a naturally occurring gas that may be present in formations and carried in hydrocarbon streams or can form as a result of sulfate-reducing bacteria (SRB) that are present in your process system. In some cases, the presence of H_2S can greatly affect the heating value of the gas, making it difficult to sell, due to strict requirements regarding the energy content of gas (Figure 6.1).

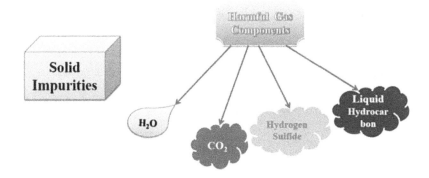

FIGURE 6.1 Components of harmful gas.

Product type and quality will affect operating conditions as well as design requirements. Waxy oils, for example, require facilities for regular on-line pigging in order to maintain the flow capacity. The control of gas quality is recommended for the following reasons:

- It helps prevent deposits lodging in the pipeline and causing a buildup of debris and the consequent pressure loss.
- It helps minimize internal corrosion caused by moisture or particular constituents such as H_2S, which causes corrosion and can assist stress corrosion.
- Hydrocarbon liquids can condense and collect in low sections of pipelines, causing a restriction to gas flow. Such liquids are also a hazard during maintenance operations, particularly if they are unexpected. For example, in the United Kingdom, all hydrocarbon liquids are subject to customs and excise control, and any such liquid recovered from a pipeline should be notified. Gas treatment plants should therefore exercise control over the water dew point, hydrocarbon dew point, and sulfur content, and these values should be monitored.

Compressors can be a source of gas contamination by releasing oil from the seals in the compressor casing. Large quantities of oil can collect in the low sections of pipelines, and the lighter droplets or fog can be carried forward into above ground installations where it blocks filters and contaminates instrumentation. Regular maintenance and performance monitoring can avoid this problem (Figure 6.2).

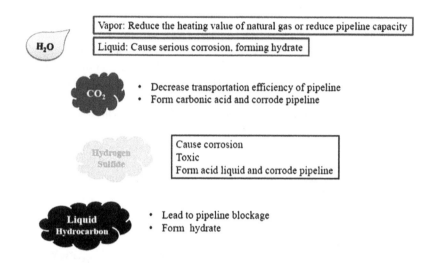

FIGURE 6.2   Effect of harmful gas components.

To mitigate the risks associated with H_2S, gas conditioning system, as described in Section 3.6.1.2.1, can help remove H_2S from your process stream, improving your production safety and meeting sales specifications.

6.2   NATURAL GAS SPECIFICATIONS

In order to ensure that natural gas supplied by one organization passes through the facilities of another operator with no technical problems, it will be essential that it should meet certain quality aspects; in other words it should conform to a particular specification such as

- Gas Safety (Management) Regulations (GS (M)R), 1996
- European Association for the Streamlining of Energy Exchange – Gas (EASEE – gas)
- ISO 13686:2013, Natural Gas – Quality Designation

Where, e.g., natural gas is produced by one or several operators from the same or different fields and is subsequently transmitted by high-pressure pipeline, gas quality will have to be defined at least in regard to:

- Water content
- Condensable content, e.g. by specifying hydrocarbon dew point at elevated pressure

- A density range
- A pressure range
- A maximum temperature

For the transfer of pipeline gas from a transmission company to a liquefaction plant, one would have to specify as a minimum:

- Water content
- Acid gases (carbon dioxide, hydrogen sulfide)
- Heavier hydrocarbons (ethane, propane, etc.)
- Nitrogen content

When selling liquefied natural gas (LNG) either into a transport chain or into storage to be ultimately regasified and distributed, seller and purchaser must agree on:

- Content of inert gases (nitrogen, helium, argon)
- Heavier hydrocarbons
- Calorific value (CV)

Finally, a gas distribution company must have assurances from its gas suppliers in regard to:

- Combustion properties
- Sulfur
- Water content

It is clear from Table 6.1 that a specification relating to the transmission and distribution of natural gas must have items relating to the integrity of both the pipeline and the flow of gas. The composition of gas entering the pipeline must be such that the pipeline is not damaged. Corrosion, stressing, and abrasion must be avoided. The composition must also ensure that, under all temperature and pressure conditions of distribution, the flow of gas can occur. Accumulation of liquid in the pipeline will reduce the capacity of the pipeline and interfere with instrumentation and control equipment. Such interference can lead to erroneous measurements and instrument failure.

The specification has the aspect covering:

- Safe use of domestic appliances
- Transmission and distribution needs
- Statutory requirements, e.g., hydrogen sulfide content

There is also a requirement that gas delivered shall be free from materials/dust, solids, or liquids which might interfere with the operation of the lines, meters, or regulators. Such gas shall be free from objectionable odors so that a distinctive "gas odor" can be added to meet the requirements of the gas safety (management regulations).

The reason that the listed items are included in the specification is given below.

TABLE 6.1
Gas Quality Specifications

Parameter	Norwegian	GS (M)R	Typical NTS (10YS)	IUK	EASEE – Gas Marcogaz
Gross CV	38.1–43.7 MJ/Sm³		36.9–42.3 MJ/Sm³	38.9–44.6 MJ/Nm³	35.01–45.18 MJ/Sm³
Wobbe index	48.3–52.8 MJ/Sm³	47.2–51.41 MJ/Sm³	48.14–51.41 MJ/Sm³	48.23–51.17 MJ/Sm³	47.0–54.0 MJ/Sm³
Oxygen	0.1 mol%	0.2 mol%	10 ppm	10 ppm	10–1000 ppm
Carbon dioxide	2.5%		2 mol%	2 mol%	2.5 mol%
Hydrogen sulfide	5 mg/Nm³	<=5 mg/m³	Max 3.3 ppm	Max 3 ppm	
Total sulfur	15 ppm	<=50 mg/m³	15 ppm	15 ppm	<=30 mg/Nm³
Water dew point	–18°C at 69 barg	See note 1	–10°C at a delivery pressure	–10°C at 69 barg	–8°C at 69 barg
Hydrocarbon dew point	–10°C at any pressure above 50 barg	See note 1	–2°C at 75 barg	–2°C at 69 barg	–2°C at any pressure above 69 barg
Delivery temperature			Between 1°C and 38°C	Between 2°C and 38°C	
Hydrogen		0.1 mol%	0.1 mol%		0.1 mol%
ICF		Less than 0.48	Less than 0.48		
Soot index		Less than 0.60	Less than 0.60		Replace with RD0.5548–0.70
Inerts			Not more than 7 mol%		
Nitrogen			Not more than 5 mol%		

Note 1: Shall be at such levels that they do not interfere with the integrity or operation of pipes or any gas appliance (within the meaning of regulation 2(1) of the 1994 Regulations) which a consumer could reasonably be expected to operate;

This transmission specification is the primary criterion in considering whether a gas needs processing. If the gas lies outside this specification, an evaluation is made of the possibility of mixing with other supplies, although this can have an effect on the security of supplies (Table 6.2).

TABLE 6.2

Natural Gas Quality Index for Pipeline Transportation of Some Developed Countries

Country	Total sulfur (mg/m^3)	H_2S (mg/m^3)	CO_2 (% V/V)	Gross heating value (MJ/m^3)
United States	23	5.7	3.0	43.6–44.3
Canada	23	5.7	2.0	36.0
Holland	50	5	2.0	38.8–42.8
France	12	5	1.5–2.0	35.2
Germany	150	7	–	37.7–46.0
Italy	120	5	–	30.2–47.2

TABLE 6.3

Gas Quality Requirements of EASEE – Gas

Parameter	Description	Units	Minimum	Maximum	Recommended implementation date
WI	Gross (superior) Wobbe index	kW/m^3	13.60	15.81	01/10/2010
D	Relative density	m^3/m^3	0.555	0.700	1/10/2010
S	Total sulfur	mg/m^3	–	30	1/10/2006
H_2S + COS	Hydrogen sulfide + carbonyl sulfide	mg/m^3	–	5	1/10/2006
RSH	Mercaptans	mg/m^3	–	6	1/10/2006
O_2	Oxygen	mol%	–	0.01[1]	1/10/2010
CO_2	Carbon dioxide	mol%	–	2.5	1/10/2006
H_2O DP	Water dew point	°C at 7000 kPa (a)	–	−8	See note 2
HC DP	Hydrocarbon dew point	°C at 1–7000 kPa (a)	–	−2	1/10/2006

[1] EASEE-gas has organized an oxygen measurement survey, which by the end of 2005 will examine the maximum feasible limit equal to or at an alternative specified value below 0.01 mol%.

[2] At certain cross-border points, less stringent values are used than defined in this CBP. For these cross-border points, these values can be maintained and the relevant producers, shippers, and transporters should examine together how the CBP value can be met in the long run. At all other cross-border points, this value can be adopted by October 1, 2006.

The European Association for the Streamlining of Energy Exchange – Gas, EASEE – Gas has proposed the gas quality specification as detailed in Table 6.3 for harmonization across the European Union. The parameter units and reference conditions are in accordance with the EASEE – Gas Common Business Practice (CBP) 2003-001/01. The energy unit is kWh, with a combustion reference temperature of 25°C. The volume unit is m³ at a reference condition of 0°C and 101.325 kPa(a). For conversion to other reference conditions, the procedures described in ISO 13443:1996 Natural Gas – Standard reference conditions should be used. The parameters are defined in ISO 14532:2001 Natural gas – Vocabulary. The specification applies only to high-calorific value (H gas) gas without added odorants.

When a gas company negotiates the purchase of a new gas supply, it is necessary to establish "how much and what it is", i.e. quality and quantity. Aspects that require clarification include:

    i. Is the composition the same for all of the gas field?
    ii. How reliable is the given composition? What variations are expected?
    iii. Are any other natural gas supplies possible? If so, when? What is known about the composition?
    iv. What processing is the seller intending? In particular, hydrocarbon dewpoint.
    v. What requirements for gas composition does the seller have? What restriction on gas composition is placed by the seller's transmission and compression requirements?
    vi. Which pressure does the seller intend for the delivery of gas?
    vii. Can the supply be guaranteed for 365 days per year?
    viii. Can maintenance schedules of the seller influence the availability of gas?
    ix. What plans does the seller have for supplying gas to other markets? What requirements do these markets place on gas composition?
    x. If hydrogen sulfide content is quoted zero, is the statement reliable? How, and for how long, was hydrogen sulfide tested for?
    xi. What is the total sulfur content of the gas?

Answers to these questions give the company confidence that the chemical and physical properties of the gas will conform to the transmission specification and gas quality statutory obligations (Nasr, Connor, and Burby, 2002).

6.2.1 Hydrocarbon Dewpoint

A temperature that is low enough to ensure that hydrocarbon liquid does not form under all temperature and pressure conditions of distribution must be specified. The dewpoint temperature specified will be governed by the lowest ambient temperature that the gas will experience, while the gas is at pressures between 24 and 38 barg (350 and 550 psig). This is the pressure range at which retrograde condensation occurs. This phenomenon must be avoided in the transmission and distribution system. If too high a hydrocarbon dewpoint is allowed, a gas which is in a single phase at high pressure can change to two phases (gas and liquid) at lower pressure even though the gas is not cooled. The extent of the two-phase region within temperature and pressure coordinates depends on the gas composition. In particular, the amount of heavy hydrocarbons can give rise to retrograde condensation. The amount of various heavy hydrocarbons allowed in the gas is inversely proportional to the carbon number of the hydrocarbon; i.e. in the series C_6, C_7, C_8, C_9, C_{10}, very much less C_{10} can give rise to retrograde condensation that of C_6. For most natural gas, after separation of the gas and condensate, the gas phase will contain too much C_{6+} component. This must be removed by gas processing, for which a chiller plant is usually employed (Table 6.4).

TABLE 6.4
Requirements for Hydrocarbon Dew Point in Different Countries

Country	Limit (°C)
ISO	There is no liquid hydrocarbon or water with the custody transfer pressure and temperature (refer to ISO 13686).
EASEE – gas	The hydrocarbon dew point is −2°C when the pressure is between 1 and 70 bar. It is issued on October 1, 2006.
Belgium	It is −3°C under 69 bar.
Argentina	−4 at 5500 kPa abs.
Holland	It is −3°C under 70 bar.
Western Australia	It varies between 0 and 10, depending on the region.
Brazil	It varies between 0 and 15 at 4500 kPa, depending on the region.
Mexico	−2 at 8000 kPa.
Peru	–
Ecuador	−2 at 8000 kPa.
Venezuela	20°C difference between gas and ambient temperature.
New Zealand	2 at 5000 kPa,
Iran	It varies between −10 and −5, depending on the gas source.
European Union	−2 at 7000 kPa.
Canada	It varies between −10 and −6.7, depending on the region.
United States	It varies between −10 and −6.7, depending on the region.
UAE	−2 at max. 6900 kPa.
Singapore	12.8 at 5000 kPa.
England	It depends on the specification: −2 up to 8500 kPa or "such that liquid does not interfere with integrity of network or appliances". Summer: It is 10°C under 69 bar. Winter: It is −1°C under 69 bar.
Austria	It is −5°C under 40 bar.
France	−2 up to 7000 kPa.
Germany	Ground temperature at pipeline operation temperature.
Italy	−10°C under 60 bar.
Colombia	7.2 at any pressure.
Russia	Temperate area: it is 0°C; summer of frigid area: it is −5°C; winter of frigid area: it is −10°C.

6.2.2   Water Dewpoint (Moisture)

The presence of liquid water in the transmission system must be avoided; otherwise hydrate formation and pipeline corrosion can occur. Hydrates are a physical combination of the lower hydrocarbons and water, and once formed they remain stable. As a result the pipeline diameter decreases, reducing the flow capacity of the line. In the extreme, blockage of the line can occur. Hydrates will also interfere with the correct operation of instrumentation, and it is more likely that the blockage of supply lines to the instruments would occur. Hydrates can occur only if free water is present; i.e. the gas is at 100% relative humidity (RH). Corrosion, however, can occur at levels below 100% RH. Protection against corrosion is ensured by operating at not more than 50% RH. The RH of gas is often conveniently expressed as a water dewpoint (Nasr, Connor, and Burby 2002).

6.2.3   Carbon Dioxide

In the early gas purchase contract of some countries, it was considered necessary to specify a maximum carbon dioxide content of 2%. This was to ensure that acid gas corrosion of the 70 bar (1000 psig) transmission system was avoided. Later it became apparent that protection against corrosion was being secured with the water dewpoint limit. It remains necessary to have gas supplies of low carbon dioxide content due to requirements at the LNG plants.

6.2.4   Oxygen

The figure of 0.01 mol% was adopted rather than 0 mol%, which in practice is the oxygen content of natural gas, to overcome measurement difficulties and false readings which can occur when endeavoring to determine a zero concentration. This 0.1 mol% limit was sufficiently low to cause air ballasting by gas sellers to be generally uneconomic. However, the molecular sieve purification units at LNG plants can be damaged by small oxygen concentrations.

6.2.5   Hydrogen Sulfide

Most gas acts require that gas shall contain a maximum of 3.3 ppm (volume/volume) hydrogen sulfide when distributed to customers. However, 3.3 ppm is seen as the limit and not the level for normal operation. Supplies are usually purchased at a much lower level.

6.2.6   Total Sulfur

This category is comprised of mercaptans, organic sulfide, and hydrogen sulfide. For example, the United Kingdom (NTS) has adopted a limit of 15 ppm (by volume) to control the amount of corrosion which could occur in domestic appliances following gas combustion. In practice, the level of total sulfur in gas delivered to the NTS is governed by the requirement that the gas is free from objectionable odor. The standard odorization of a smell-free gas introduces 5 ppm (by volume) of total sulfur (Nasr, Connor, and Burby 2002).

6.2.7 TEMPERATURE

Too low a delivery temperature may lead to freezing of the soil around a buried pipeline, and damage to other pipelines and services can result. Conversely, a temperature too high can be injurious to pipeline wrappings and coatings. The temperature range of the transmission specification reflects these requirements.

6.2.8 GAS INTERCHANGEABILITY

When formulating a gas specification, it is essential to take account of how the gas will burn on domestic appliances and the compatibility of the gas and the range of appliances. The International Gas Union (IGU) recognized the need to categorize gases according to their properties. Three families of gas were identified:

1) Family 1 – gases having a Wobbe number in the range 22.4–30.0 MJ m^{-3} (st) (607–800 Btu ft^{-3}(st) (dry)). This family embraces the manufactured town gases of various types and LPG/air mixtures within the same Wobbe number range.
2) Family 2 – gases having a Wobbe number in the range 39–55 MJ m^{-3} (st) (1056–1472 Btu ft^{-3} (dry)). This family was intended to include all the various natural gases, including synthetic gases and some LPG/air mixtures.
3) Family 3 – gases having a Wobbe number in the range 73–88 MJ/m^3 (st) (1972–2353 Btu/ft^3 (dry)). This family consists of liquefied petroleum gases (LPGs), i.e. essentially propane and butane.

The variety of gases within each family is too great, however, for all the gases to be fully interchangeable without appliance adjustment, and it has been necessary to subdivide each family into "groups". For natural gases, the "second family" was divided into two groups: Group L (based on the Dutch natural gas) consisting of gases in the lower part of the Wobbe number range (39–45) and Group H consisting of gases of higher Wobbe number (45.7–55 British natural gas). For each group a "reference" or "adjustment" gas was defined, together with limit gases, for use in the testing of appliances.

Within each group gases which have Wobbe numbers within +5% and −5% of the reference Wobbe number should be fully interchangeable, i.e. burn safely, cleanly and efficiently without a need for appliance adjustment. This 5% guideline holds good, while gases have compositions not markedly different from the reference gas. However, for group H gas the methane content of the gas must not be less than 80% and 85% for the Wobbe index alone to be a reliable guide to combustion characteristics. Also, a normal distribution limit of plus 3% of reference Wobbe number was adopted. However, this aspect is also covered in the Gas Safety (Management) Regulation – Schedule 3.

6.2.9 CALORIFIC VALUE

The most important property a fuel gas possesses is the energy liberated when it is burned. This may be expressed as the heats of formation of its combustion products on a molar basis. In the fuel industry, however, this property is much more commonly expressed as the calorific value (CV), which is the quantity of heat released by complete combustion under isothermal conditions at a constant pressure of one atmosphere and at a specified reference temperature of a unit quantity of the fuel, the water formed during the combustion being in the liquid state, any sulfur in the fuel being converted to sulfur dioxide, and any nitrogen remaining as such. Calorific value is also known as heating value.

Complete combustion is possible with gaseous fuels at atmospheric pressure, and the CV is measured at a constant pressure in a calorimeter. Solid and liquid fuels, on the other hand, require higher pressure and determinations of CV are made under constant volume conditions in a bomb calorimeter.

Fuel gases which contain hydrogen or hydrocarbons possess two CVs: the superior (gross) CV and the inferior (net) CV, depending upon whether the water formed in combustion is in the liquid or vapor phase (Nasr, Connor, and Burby 2002).

The superior (gross) calorific value of the gas (relative to the volume of the dry gas) is defined as the amount of heat given out by the complete combustion of the gas with air, at a constant pressure of 1.01325 bar and a constant temperature t_H of a specific volume (V) under specified conditions (t_v, P_v), all the water that is formed during the combustion being condensed at the temperature t_H. The superior CV is designated:

$$H_s\{t_H, V(t_v, P_v)\} \qquad (6.1)$$

The inferior (net) calorific value of the gas (relative to the volume of dry gas) is defined as the amount of heat given out by the complete combustion of the gas with air, at a constant pressure of 1.01325 bar and a constant temperature t_H of a specific volume (V) under specified conditions (t_v, P_v), all the water that is formed during combustion remaining in the gaseous phase at the temperature t_H. The inferior calorific value is designated:

$$H_i\{t_H, V(t_v, P_v)\} \qquad (6.2)$$

The gross CV provides the basis on which charges are made by the gas industry on their consumers. The specified conditions referred to above on the International System of Units are known as "standard reference conditions" (src) or "metric standard conditions" (MSC): a temperature of 15°C and a pressure of 101325 Pa (dry). On this system the gross CV is expressed in MJ/m³ (st).

The determination of the CV for the purpose of selling gas to be used for heating (rather than chemical feedstock) requires a high accuracy calculation based on a high-quality gas chromatographic analysis. This method can be as reliable as a direct measurement of CV in a calorimeter.

Using ISO 6976 method, the ideal calorific value H is calculated from:

$$H_{(ideal)} = \sum_{j=1}^{j=n} X_j H_{j(ideal)} \tag{6.3}$$

where $H_{(ideal)}$ = ideal calorific value of component j in the mixture and $X_{j(ideal)}$ is the mole fraction of component j. The method applied to both gross and net calorific values. The real calorific value $H_{(x)}$can be obtained from the ideal value as follows:

$$H_{(real)} = \sum_{j=1}^{j=n} \frac{X_j H_{j(ideal)}}{Z} \tag{6.4}$$

where Z is the compressibility factor of the mixture at standard conditions.

6.2.10 RELATIVE DENSITY

The relative density of gas (d) is the ratio of the density of the gas to the density of air at standard conditions.

It may also be defined as the gas quality contained in a certain volume divided by the quality of dry air under the same reference condition. In other words, it is a ratio of the density of natural gas (ρ_{gas}) to the density of dry air (ρ_{air}) under the same reference.

$$d_{(ideal)} = \frac{\rho_{gas}}{\rho_{air}} = \frac{M_{gas}}{M_{air}} \tag{6.5}$$

and

$$d_{(real)} = \frac{\rho_{gas} \times Z_{air}}{\rho_{air} \times Z_{gas}} = \frac{M_{gas} \times Z_{air}}{M_{air} \times Z_{gas}} \tag{6.6}$$

For most practical purposes the difference between the real and ideal values at standard conditions is negligible but can be important for the determination of CV and for gas flow measurement.

However, for combustion calculations it is often permissible to calculate the calorific value and relative density of a multi-component fuel gas by a simple additive method from the CVs and relative densities of its constituents assuming ideal gas behavior. This is because at the low partial pressures of the constituents, the compressibility factors are very close to unity and a sufficiently accurate result can be obtained from the ideal gas values and volume fractions.

6.2.11 WOBBE NUMBER

The heat contained in the gas is proportional to:

$$(CV) / \sqrt{d} \tag{6.7}$$

where CV is the calorific value, and "d" is the relative density of the gas.

Equation (6.7) is termed the Wobbe index or Wobbe number, and it gives a measure of the relative heat input to a burner at a fixed gas pressure of any fuel gas.

When a gas composition conforms to the specified Wobbe number and calorific value ranges and the impurity levels, it is suitable for supply to consumers. Extensive research has shown that the Wobbe number is a very good measure of the burning character.

What are the combustion consequences of the Wobbe number changes? Consider a gas supply which has a high methane content and gives satisfactory combustion, i.e. keen flames and a low carbon monoxide output. Then increase the ethane and propane content in the gas, which increases the Wobbe number. The primary air entrained remains the same, but the air requirement for complete combustion has now increased; i.e. the primary percentage aeration achieved has fallen. This results in more air being required through diffusion at the flame to complete combustion. The flame therefore lengthens and is more likely to impinge on cold heat exchanger surfaces. Laboratory work has shown a direct relationship between the Wobbe number and carbon monoxide (CO) production. Raising the Wobbe number by 1.5 MJ/m^3 (40 Btu/ft^3) will double CO output. Too great a reduction in the Wobbe number will lead firstly to a loss of heat service, and a further reduction will lead to flame stability problems; i.e. the flame can lift away from the burner (Nasr, Connor, and Burby 2002).

6.2.12   RELATIVE HUMIDITY

It is the ratio of the amount of water vapor in the natural gas at a specific temperature to the maximum amount that the natural gas can hold at that temperature, expressed as a percentage.

6.2.13   COMPRESSIBILITY FACTOR (Z)

It is also known as the compression factor. It is the ratio of the molar volume of a gas to the molar volume of an ideal gas at the same temperature and pressure.

The compressibility factor is defined as:

$$Z = \frac{V_m}{\left(V_m\right)_{ideal\ gas}} = \frac{pV_m}{RT} \tag{6.8}$$

where V_m is the molar volume, $(V_m)_{ideal\ gas} = RT/P$ is the molar volume of the corresponding ideal gas, P, is the pressure, T, is the temperature, and R is the gas constant. For engineering applications, it is frequently expressed as:

$$Z = \frac{P}{\rho R_{specific} T'} \tag{6.9}$$

where ρ is the density of the gas and $R_{specific} = R/M$ is the specific gas constant, M being the molar mass.

For an ideal gas the compressibility factor is $Z = 1$ per definition. In many real-world applications requirements for accuracy demand that deviations from ideal gas behavior, i.e. real gas behavior, be taken into account. The value of Z generally increases with pressure and decreases with temperature. At high pressures molecules are colliding more often. This allows repulsive forces between molecules to have a noticeable effect, making the molar volume of the real gas (V_m) greater than the molar volume of the corresponding ideal gas (($V_m)_{ideal\ gas} = RT/P$), which causes Z to exceed one. When pressures are lower, the molecules are free to move. In this case attractive forces dominate, making $Z < 1$. The closer the gas is to its critical point or its boiling point, the more Z deviates from the ideal case.

6.2.13.1 Compressibility Factors at Standard Conditions

For most gas engineering problems, the difference between ideal gases and real gases at 15°C and 1.01325 bar can be ignored; however, where high accuracy is required the compressibility factor at standard conditions should be used. The most important areas where this will be required are where the monetary value of gas is involved. Typical examples are the calculation of calorific value as the basis of sale for the gas and custody transfer metering which usually involves large quantities of gas being transferred between different utilities or being sold to industrial customers.

The basis for calculating the compressibility factor of a gas mix at standard conditions is given in ISO 6976 as:

$$Z_{mix} = 1 - \left(\sum y_i / \sqrt{b_i} \right)^2 + 0.0005 \left(2Y_H - Y_H^2 \right) \tag{6.10}$$

where Y_H is the mole fraction of hydrogen in the mixture

b is the gas law deviation coefficient $= 1 - Z$

Values for the compressibility factors of single gases at standard conditions are given in the standard and should be used to calculate the value of Z_{mix}, which is then used to convert ideal gas properties to real gas properties at standard conditions.

The decision to use this equation will depend on the accuracy of the other data involved in the particular calculation. As an example the compressibility factor for methane at standard conditions is 0.998; therefore other data involved in a calculation with methane should be accurate to at least the same order, i.e. 0.2%.

The American Gas Association has produced tables of "super compressibility factors", which they define as:

$$F_{PV} = 1 / \sqrt{Z} \tag{6.11}$$

The values quoted in the tables are for the value F_{PV}/F_{PV°. F_{PV° is the super compressibility factor at standard conditions (60°F and 14.7 lbf/in²) and are intended for use as a correction factor to be applied to high-pressure metering rather than in the more general engineering calculations. The term "super compressibility" should be used with caution since it is often confused with compressibility factor (Nasr, Connor, and Burby 2002).

6.3 ADJUSTMENT OF GAS QUALITY

For gas that does not meet the specified gas quality regulation or specifications, I recommend the following measures should be accepted to assure the integrity of pipelines.

1. Blending

    Gas quality adjustment within the network, known as gas mixing or blending. Gas blending is the process of mixing gases for a specific purpose where the composition of the resulting mixture is specified and controlled.

- Within networks not possible if risk of off-specification gas to customers. However, in some circumstances gas mixing in networks is possible and may enable a rich gas to be diluted or a lean gas to be enriched to meet a supply specification. Gas mixing within the network may be possible provided that continuous gas supplies of appropriate quality are available.

  For example, if a gas from the Barrow terminal, situated on the west coast of England, is very lean with a Wobbe Index below the GS(M)R limit. By mixing the Barrow gas with the richer gas flowing through the network from the North Sea fields within an underground loop of pipeline specifically constructed for this purpose, the Wobbe Index is raised to an acceptable level.

  Gas mixing within the network could be used for:
  - Separate gas supplies, of which one or more may be out of specification.
  - Out-of-specification LNG boil-off gas with pipeline gas.
  - Export flows from different LNG tanks to minimize the propane enrichment.
  - LNG send-out alongside pipeline gas.

  In older gas networks, cast iron pipes may be joined with lead and yarn that need to be kept swollen to maintain the seal. The introduction of relatively dry gas including regasified LNG could be implicated in deterioration of the seal, resulting in leaks and compromising safety.
- At beach sub-terminal possible availability/security of supply

2. Gas treatment
   a. Derichment of high Wobbe index gas (i.e. lowering the Wobbe index)
      i. Natural gas liquids (NGL) removal possibly with reforming.
         - Side stream reforming: catalytic conversion of side stream into lower CV gas. Moderate complexity, previously used in Italy.
      ii. Cryogenic distillation: removal of heavier hydrocarbons from LNG. Moderate complexity, previously used in Spain. Needs LPG sales capability.
      iii. Ballasting with inert gas
         - Nitrogen (N_2) injection: ballasting of high Wobbe index gas with nitrogen to reduce the calorific value. Large volumes of nitrogen (N_2) are required. It is a simple technology, has widespread use, and can be outsourced.
         - Most LNG is too rich for the market. Ballasting with nitrogen brings within specification.
   b. Enrichment of low Wobbe index gas
      i. LPG injection
      ii. Inerts removal
3. Change of specification
4. Curtailment or cessation of flows of gas with characteristics outside gas quality specification

Note:

Wobbe index/number is widely regarded as the best single measure of the interchangeability of fuel gas.

6.4   PROPERTIES RELEVANT TO LIQUEFACTION

Another important set of gas characteristics are those which have a bearing on natural gas liquefaction. Since the volume of methane can be reduced by a factor of about 600 by cooling and compressing the gas until it turns into a liquid, there is considerable incentive to ship, store, and possibility even distribute it in liquid form. As shown in Table 6.5, however, this is not an easy matter; the boiling point of methane is −162°C, and such low temperatures are difficult to achieve; clearly the liquefaction of ethane, propane, and butane are far easier. Importantly also the critical temperature of methane, i.e. the temperature above which one cannot liquefy it simply by the compression of the gas, is low (−83°C), and methane must be precooled to that level before the ultimate liquefaction pressure is applied (Nasr, Connor, and Burby 2002).

TABLE 6.5
Properties Relevant to Liquefaction

	Methane	Ethane	Propane	Butanes
Boiling point (°C)	−162	−89	−42	−1 to −12
Critical temperature (°C)	−83	32	97	135 to 152

6.5   THE MORE IMPORTANT GAS CHARACTERISTICS

Natural gases are produced, purified, compressed, shipped, stored in compressed or liquid form, distributed, and eventually burned or converted into chemicals. Table 6.6 lists gas characteristics which are of particular significance for each operation (Nasr, Connor, and Burby 2002).

TABLE 6.6
Gas Characteristics of Importance for Each Operation

	Production	Liquefaction	Compression	Transport	Storage	Combustion
Relative density			X	X		
Dew point (hydrocarbon)	X	X	X	X	X	
Boiling point		X			X	
Acid gas content	X	X	X	X	X	
Sulfur content	X	X				X
Nitrogen content		X				X
Water content	X	X	X	X		
Combustion characteristics						X

7 Natural Gas Hydrate

7.1 INTRODUCTION

An early part of the design of gas processing facilities and gas transmission pipelines is the prediction of hydrocarbon hydrate formation. Hydrocarbon hydrates are solid compounds which form at well-defined conditions of temperature and pressure in the presence of free water.

Natural gas hydrates (*gas hydrates*) are crystalline water-based solids physically resembling ice in which a hydrocarbon, usually *methane*, is trapped (Figure 7.1). A wide range of molecules form gas hydrates, and those of most practical interest are light hydrocarbons such as methane, ethane, and propane. They occur in the pore spaces of sediments and may form cements, nodes, or layers. Carbon dioxide and hydrogen sulfide also form hydrates and are of particular interest.

FIGURE 7.1   Typical example of natural gas hydrate.

Formation of gas hydrates may give rise to operational problems such as heat exchanger tube blocking, instrumentation plugging, or pipeline blocking; hence the conditions suitable for their formation should be avoided whenever possible. Their accumulation in processing facilities or pipelines restricts or stops the flow of fluids, causing shutdowns and even destruction of valuable equipment (see Figure 7.2).

The purpose of this chapter is to present the procedures and reference materials to be used in the prediction of conditions suitable for hydrocarbon hydrate formation. Additionally, this chapter describes the methods of preventing hydrocarbon hydrate formation and methods for removing them in the event of their formation.

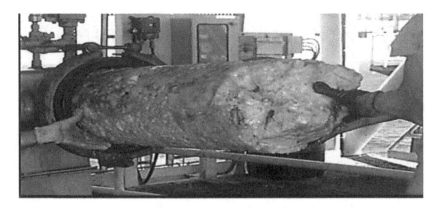

FIGURE 7.2 Hydrates are blocking the pipeline.

7.2 STRUCTURE OF GAS HYDRATE

In the production and processing of fluids containing hydrocarbons and water, there is a risk of hydrate formation. Natural gas hydrates are ice-like crystalline compounds formed when natural gas components such as methane, ethane, propane, isobutane, hydrogen sulfide, carbon dioxide, and nitrogen are entrapped in a crystal lattice of water molecules. The water cage-like structure is called "host", and the entrapped gas molecules are called "guest" (see Figure 7.3).

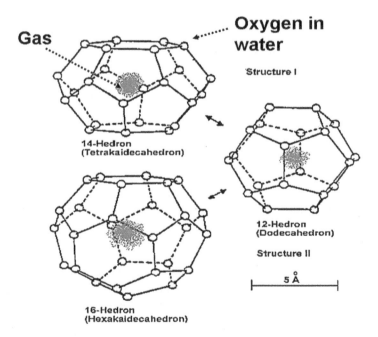

FIGURE 7.3 Structure of gas hydrate (Source: www.sps.esd.ornl.gov/desalinationpage. html).

The individual gas hydrate crystals consist of water and the molecules of one or several hydrate-forming gases. The individual molecules of these gases become attached within cavities in the water molecular structure and form a crystalline lattice which is stronger than the original water compound and therefore is insoluble in water. However, the hydrocarbon molecule does not form any stable chemical bond with the water molecules.

The resulting hydrate crystals have, in the main, the form of one or two pentagonal dodecahedra and can be described by the following structural formula:

$$C_nH_{2n+2} \cdot xH_2O \tag{7.1}$$

where

x = the coordination number and is usually equal to 6 or 7.

Hydrocarbon hydrates may be formed by methane, ethane, propane, n-butane, or isobutane. Larger hydrocarbon components cannot form hydrates as their molecules are too large. Other components of natural gas, such as nitrogen, carbon dioxide, and hydrogen sulfide may form solid hydrate compounds.

7.2.1 TYPES OF HYDRATES

Three types of gas hydrates have been identified:

- Type I: the guest molecules are small gas molecules (methane, ethane, H_2S, CO_2).
- Type II: the guest molecules are larger gas molecules (propane, isobutane, and n-butane).
- Type III: the guest molecules include benzene, cyclopentane, and cyclohexane.

7.3 CONDITIONS NECESSARY FOR HYDRATE FORMATION

The conditions at which hydrate formation is most likely to occur correspond to high pressure and low temperature. For pipeline systems the lowest temperature normally occurs during shut down; hence this situation must always be considered. Free water must be present within the pipeline or processing system.

Therefore, conditions which are known to promote hydrate formation are:

a. Low temperature
   - Hydrates form at a low temperature called formation temperature. If a gas containing water is cooled below its hydrate formation temperature, hydrates will form.
b. High pressure
   - High pressure favors hydrate formation.
c. Water
   - Free water or water vapour at sufficiently low temperature or high pressure enhances hydrate formation. No hydrate formation is possible if

"free" water is not present. Here, we understand the importance of removal of water vapor from natural gas, so that in case of free water occurrence there is likelihood of hydrate formation.

Further conditions that are known to promote hydrate formation are

- High velocities
- Pressure pulsations (in other words, turbulence can serve as a catalyst)
- Agitation
- **Presence of H$_2$S and CO$_2$** promotes hydrate formation because both these acid gases are more soluble in water than the hydrocarbons.

It is important to note that:

- High BTU gas is more likely to produce hydrates and freezing problems.
- Joule–Thomson rule of temperature effect because of pressure reduction: temperature will decrease by approximately 7°F for every 100-psi pressure reduction.

The hydrate crystals forming on pipe walls initiate any unevenness such as weld seams or areas of corrosion. Narrowing of the internal pipe diameter starts slowly, but as flow and pressure conditions change, hydrate build-up is accelerated. Natural gas hydrate (NGH) will easily form at:

- Elbow
- Weld seams
- Areas of corrosion
- At the downstream of the valve where is a big pressure drop
- Relief valve and the safety valve of the high-pressure vessel
- Pig launcher and pig receiver

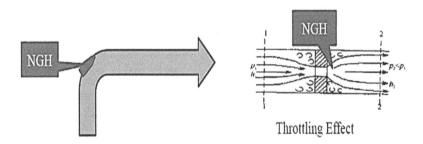

Throttling Effect

FIGURE 7.4   Places where the hydrate will form easily.

7.4 PREDICTION OF HYDRATE FORMATION CONDITION

The prediction of temperature and pressures at which hydrates will form may be estimated from the hydrocarbon gas gravity or calculated more accurately from the gas composition (Figure 7.4).

7.4.1 PREDICTION FROM GAS GRAVITY

The gas gravity method is the simplest method for quantifying the hydrate formation temperature and pressure. Gas gravity is defined as the molecular weight of the gas divided by that of air. To use the chart shown in Figure 7.5, calculate the gas gravity and specify the pressure/lowest temperature of the pipeline/process. The pressure/temperature at which hydrates will form then is read directly from the chart at that gas gravity and pressure/temperature.

Figure 7.5 shows the hydrate formation temperatures and pressures for natural gases of varying gravity. The effect of nitrogen on the hydrate formation conditions is also shown. Any quantity of nitrogen in the natural gas tends to decrease the temperature, for a constant pressure, at which hydrates will form.

This correlation is not suitable for the prediction of hydrocarbon hydrate formation conditions for natural gas streams containing hydrogen sulfide.

An example of hydrate formation prediction using Figure 7.5 is shown below.

7.4.1.1 Example Calculation

1. Calculate the temperature at which a natural gas stream at a pressure of 2000 kPa or 290.08 Psia will form hydrates. The natural gas stream has a molecular weight of 20.00 (the molecular weight of air is 28.966).
   Solution

$$\gamma_g = \frac{\overline{M_g}}{M_a} \tag{7.2}$$

where

$\overline{M_g}$ = the total molecular weight of the gas in the mixture
M_a = the molecular weight of air

$$\text{Gas gravity}\left(\gamma_g\right) = \frac{\overline{M_g}}{M_a} = \frac{20}{28.966} = 0.69$$

From Figure 7.5,
Hydrate formation temperature = 51.8°F = 11°C

2. A gas is composed of the following (in mole percent):
   92.67% methane
   5.29% ethane
   1.38% propane
   0.182% isobutane
   0.338% n-butane
   0.14% pentane

When free water is present with the gas, find:
The pressure at which hydrates form at 283.2 K (50°F).

Solution

TABLE 7.1
Calculation of Average Molecular Weight

Component	Mole fraction in mixture, z	Molecular weight, M	Average M of gas in mixture, Mg $Mg = z \times M$
Methane	0.9267	16.043	14.867
Ethane	0.0529	30.070	1.591
Propane	0.0138	44.097	0.609
Isobutane	0.00182	58.124	0.106
n-Butane	0.00338	58.124	0.196
Pentane	0.0014	72.151	0.101
Total	1.000		17.470

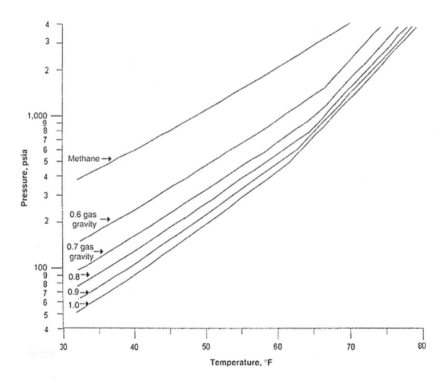

FIGURE 7.5   Hydrate formation conditions for gases of varying gravity (air = 1) (GPSA 1980).

The gas gravity (γ_g) is calculated as 0.603, using the average molecular weight calculated in Table 7.1 and Equation (7.2):

$$\gamma_g = \frac{\overline{M_g}}{M_a} = \frac{17.470}{28.966} = 0.603$$

where

$\overline{M_g}$ = the total molecular weight of the gas in the mixture
M_a = the molecular weight of air

Using this gas gravity number to read Figure 7.5 indicates that:At 50°F, the hydrate formation pressure is 450 psia at a gas gravity of 0.603.

7.4.2 PREDICTION USING VAPOR–SOLID EQUILIBRIUM CONSTANTS

Knowledge of the temperature and pressure of a gas stream at the wellhead is important for determining whether hydrate formation can be expected when the gas is expanded into the flow lines. If the composition of the stream is known, the hydrate temperature can be predicted using vapor–solid (hydrate) equilibrium constants. The basic equation for this prediction is:

$$\sum \frac{Y_n}{K_n} = 1.0 \qquad (7.3)$$

where

Y_n = mole fraction of hydrocarbon component n in the gas on a water-free basis
K_n = vapor–solid equilibrium constant for hydrocarbon component n

Equation (7.4) may be solved by a trial-and-error approach. For a constant pressure, hydrate formation temperatures may be guessed (assumed), or for a constant temperature, hydrate formation pressures may be guessed (assumed).

The vapor–solid equilibrium constant is determined experimentally and is defined as the ratio of the mole fraction of the hydrocarbon component in the gas on a water-free basis to the mole fraction of the hydrocarbon component in the solid on a water-free basis. That is:

$$K_n = Y_n / X_n \qquad (7.4)$$

where:

X_n = mole fraction of hydrocarbon component n in the solid on a water-free basis

Graphs showing the vapor–solid equilibrium constants at various temperatures and pressures are given in Figures 7.6–7.10.

For nitrogen and components heavier than butane, the equilibrium constant is taken as infinity. Low concentrations of n-butane (< 5 mole%) may be represented using the equilibrium constants for ethane. For hydrocarbons heavier than butane, the equilibrium constants are taken as infinity, because these molecules are too large to form hydrates.

The steps for determining the hydrate temperature at a given system pressure are as follows:

a. Assume a hydrate formation temperature
b. Determine K_n for each component
c. Calculate Y_n/K_n for each component
d. Sum the values of Y_n/K_n
e. Repeat steps 1–4 for additional assumed temperatures until the summation of Y_n/K_n is equal to 1.0

An example of hydrate formation prediction using vapor–solid equilibrium constants is shown below.

7.4.2.1   Example Calculation

Calculate the temperature at which a natural gas stream as given below (molecular wt = 20.0) at a pressure of 2000 kPa will form hydrates.

	Mole fraction in gas (dry basis)	At 10°C		At 15°C	
Component	Y	K	Y/K	K	Y/K
Methane	0.818	2.05	0.399	2.18	0.375
Ethane	0.110	0.82	0.134	1.45	0.076
Propane	0.025	0.12	0.208	0.72	0.035
Isobutane	0.022	0.046	0.478	0.20	0.110
n-Butane	0.010	0.82	0.012	1.45	0.007
Carbon dioxide	0.015	3.0	0.005	10.0	0.002
Total			1.236		0.605

Solution

$$\text{Interpolating linearly} = 10 + \left(\frac{1.236 - 1.0}{1.236 - 0.605} \right) \times 5 = 11.9°C$$

Interpolating linearly, Y/K = 1.0 at 11.9°C; therefore, hydrate formation temperature is 11.9°C.

7.4.3   PREDICT GAS HYDRATE FORMATION TEMPERATURE
WITH A SIMPLE CORRELATION

Hydrate formation temperature (HFT) can be precisely predicted using a new, simple correlation. The proposed equation has been developed based on 22 data points, covering gas specific gravities from 0.55 to 1, and it has been compared to several well-known and accurate gravity models. Some of the most important and well-known hydrate formation correlations are reviewed in the following section.

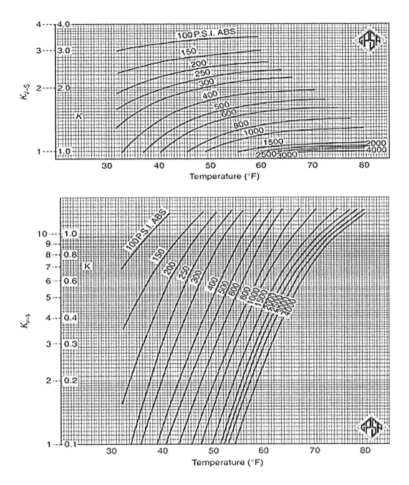

FIGURE 7.6 Vapor–solid equilibrium constants for methane (top) and ethane (bottom) (GPSA, 2004).

In 1934, Hammerschmidt proposed a correlation for gas hydrate formation, shown in Equation (7.5).

$$T_{(°F)} = 8.9 P_{(psi)}^{0.285} \qquad (7.5)$$

Where T and P, are temperature and pressure of hydrate formation, respectively. This easy-to-use equation does not take into account the effect of gas specific gravity.

However, some P-explicit hydrate correlations have also been proposed. In 1981, a famous P-explicit correlation was presented by Makogon and developed later by Elgibaly and Elkamel. A modified form of the Makogon correlation is presented in Equation (7.6).

$$\log P_{(MPa)} = \beta + 0.0497 \left(T_{(°C)} + k T_{(°C)}^2 \right) - 1 \qquad (7.6)$$

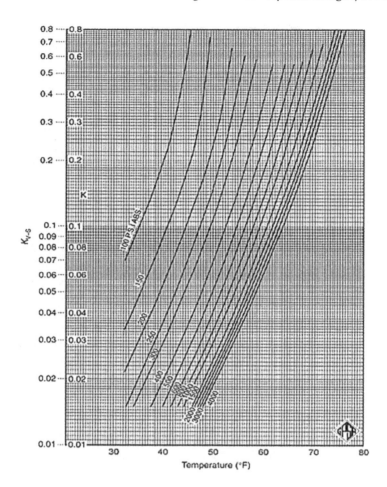

FIGURE 7.7   Vapor–solid equilibrium constants for propane (GPSA 2004).

Where:

$$\beta = 2.681 - 3.811\gamma + 1.679\gamma^2$$
$$k = -0.006 + 0.011\gamma + 0.011\gamma^2$$
$$\gamma = \text{gas specific gravity}$$

In 1991, Motiee suggested Equation (7.7):

$$T_{(°F)} = -238.24469 + 78.99667 \log P_{(Psi)} - 5.352544 \left(\log P_{(Psi)} \right)^2$$
$$+ 349.473877\gamma - 150.85467\gamma^2 - 27.604065\gamma \log P_{(Psi)} \tag{7.7}$$

Equation (7.7) is well known and widely used in the oil and gas industry because of its accuracy for natural gas mixtures.

In 2005, Towler and Mokhatab recommended a relatively simple correlation for predicting HFT of natural gas mixtures. A modified form of this correlation is shown in Equation (7.8):

$$T_{(°F)} = 13.47 \ln P_{(Psi)} + 34.27 \ln \gamma - 1.675 \ln \gamma \ln P_{(Psi)} - 20.35 \qquad (7.8)$$

In 2009, Bahadori and Vuthaluru presented a complicated and accurate correlation for estimating HFT. The general form of their calculation is shown in Equation (7.9):

$$\ln T_{(K)} = \sum_{j=1}^{j=4} A_j \left(\frac{1}{P_{(kPa)}} \right)^{j-1} \qquad (7.9)$$

where A_j is calculated as shown in Equation (7.10):

$$A_j = \sum_{i=1}^{i=4} B_{ji} \left(MW_{gas} \right)^{i-1} \qquad (7.10)$$

Equation (7.9) has sixteen (16) adjustable parameters. Implementing different sets of adjustable parameters, for a total of forty-eight (48) parameters, for different ranges of pressure and gas molecular weight (MW) is recommended (Mani, Namvaran 2015).

7.4.4 COMMERCIAL COMPUTER PROGRAMS

Commercial computer programs are the most accurate prediction of hydrate formation conditions. These programs are generally quite good, and so simple to use, they often require less time than the simplified methods presented. There are two types of commercial computer programs used in predicting hydrate formation temperature.

a. Incipient hydrate formation programs (enables the prediction of pressure and temperature at which hydrate begins to form)
b. Flash programs or Gibbs energy minimization programs (predicts all phases and amount at higher pressures and lower temperatures than the incipient hydrate formation point)

The flash/Gibbs program is gaining pre-eminence. The basis for both program types is a hydrate equation of state (EOS). When compiling critical prediction results for a design, however, it is important to verify the program results by hand calculation to determine whether the program has made an unusual prediction (see Section 7.4.1–7.4.3).

7.5 WATER CONTENT OF A NATURAL GAS STREAM

Raw natural gas streams are often saturated with water; i.e. the natural gas stream contains as much water as possible in the vapor phase. Water content is measured in natural gas pipelines at production and gathering sites, custody transfer points, compression stations, storage facilities and in the distribution markets. Any reduction of the saturated natural gas stream temperature, at a constant pressure, will result in the dropout of liquid water.

This reduction in temperature may be encountered in numerous situations, e.g. gas transmission pipelines or processing heat exchanges. Example calculation of the quantity of water dropout in natural gas pipeline using Figure 7.11 is shown in Section 7.5.1.

Free water is undesirable in gas pipelines for numerous reasons:

1. Hydrocarbon hydrates may form in the pipeline. Narrowing of the pipeline starts slowly, but as it changes flow and pressure conditions, the hydrate build-up accelerates, giving rise to larger and larger flow restrictions in the pipeline.
2. Free water dissolves acid gases in the natural gas stream, forming a corrosive fluid. Both hydrogen sulfide and carbon dioxide readily dissolve in water to form a highly corrosive fluid. Neither hydrogen sulfide nor carbon dioxide is corrosive in dry natural gas streams.
3. Sales gas contracts often specify the maximum water vapor allowable in natural gas products. For these reasons water is usually removed from natural gas streams as soon as possible, i.e. at the facility which collects gas from the numerous field wells.

7.5.1 EXAMPLE CALCULATION OF WATER DROPOUT IN A NATURAL GAS TRANSMISSION PIPELINE

Calculate the quantity of water dropout that occurs in a raw natural gas transmission pipeline with a capacity of 350 mmscf. The saturated natural gas having a molecular weight of 20 is fed into the pipeline at 1500 psia and 50°C. The natural gas arrives at the processing platform at 1400 psia and 4°C.

Solution
Water vapor content of natural gas at 1500 psia and 50°C = 75 lb/mmscf (Figure 7.11)
   Correction factor for molecular weight of 20 and temperature of 50°C = 0.99 (Figure 7.11)
   Water vapor content of natural gas at 1400 psia and 4°C = 7.3 lb/mmscf (Figure 7.11)
   Correction factor for molecular weight of 20 and temperature of 4°C = 1.00 (Figure 7.11)

$$\text{Quantity of water drop} = \frac{350\left(75\times0.99-7.3\times1.00\right)\text{lb}}{\text{day}}$$

$$= 3560 \text{ lb/day}$$

$$\therefore \text{Quantity of water drop} = 1615 \text{ kg/day}$$

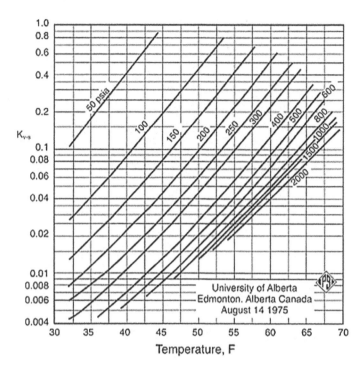

FIGURE 7.8 Vapor–solid equilibrium constants for isobutane (GPSA 2004).

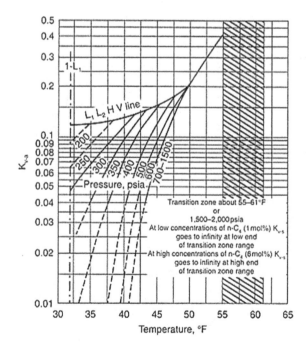

FIGURE 7.9 Vapor–solid equilibrium constants for n-butane (GPSA 2004).

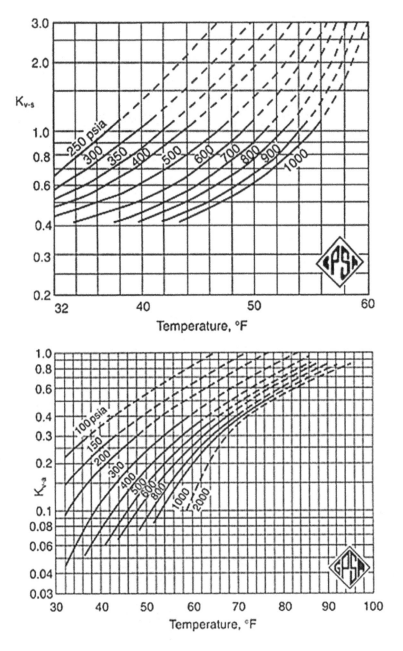

FIGURE 7.10 Vapor–solid equilibrium constants for carbon dioxide (top) and hydrogen sulfide (bottom) (GPSA, 2004).

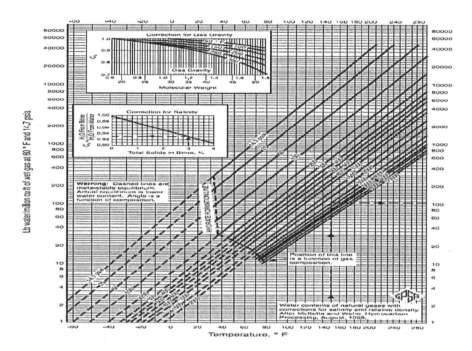

FIGURE 7.11   Dew point of natural gas (GPSA 2004).

7.6   METHODS OF HYDRATE PREVENTION AND MITIGATION

To prevent the formation of hydrates within the process the following methods are used:

1. Increasing the gas temperature above the temperature indicated on the hydrate formation curve for the given operating pressure. The temperature is maintained above the hydrate formation temperature of the gas.
2. Decreasing the pressure below the pressure indicated on the hydrate formation curve for the given operating temperature. The pressure is maintained below the value necessary for hydrates to form.
3. Following recommended chemical injection rates. Chemicals called hydrate inhibitors (methanol and/or one of the glycols) are injected into the process where hydrates are likely to form. They push the hydrate formation equilibrium to lower temperatures or higher pressures. Certain inhibitors called low-dosage inhibitors prevent any hydrate formed from coagulating and growing or delay the hydrate formation process.
4. Eliminating free water in the gas stream by dehydrating the gas or elevating the temperature to vaporize more water. Water is removed from the gas stream through the process called gas dehydration.
5. Re-designing piping systems (e.g. low points, restrictions).

Methanol, or one of the glycols, mixes with the condensed aqueous phase to lower its hydrate formation temperature at the given pressure (or increase the hydrate formation pressure at a given temperature) when injected into a gas process stream. Because these glycols shift the equilibrium to lower temperatures and higher pressures, they are termed thermodynamic inhibitors. They also lower the freezing point of the liquid water (see Figure 7.12). To be effective, the chemical inhibitor is injected at the very points where the wet gas is cooled to its hydrate temperature. The injection must be in such a way that there is good mixing with the gas stream. Both the glycols and the methanol can be recovered together in the aqueous phase and then regenerated and recycled. The choice of inhibitors depends on operating conditions and economics.

We shall look at hydrate inhibition methods and gas dehydration.

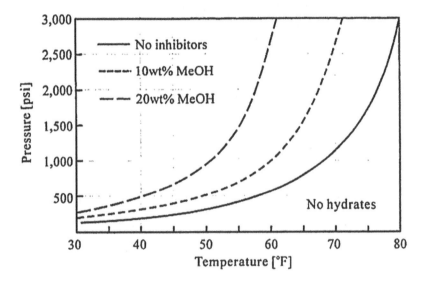

FIGURE 7.12 Effects of methanol on hydrate formation (Source: Subsea pipeline and risers by Yong B. and Qiang B. 2005).

7.6.1 Hydrate Inhibitors

Hydrate formation along a long natural gas pipeline has recently been established to initiate different types of internal corrosion along the pipe length based on the formation stage and point. These corrosions may lead to the disintegration of the pipe's properties and eventually result in the pipe's leakage or full-bore rupture. Apart from the enormous economic implications on the operating company, the conveyed fluid upon escape to the environment poses the risk of fire, reduction of air quality, and other health hazards.

Hydrate inhibitors are used to inhibit hydrate formation, e.g. methanol. Formation of hydrates may give rise to operational problems such as heat exchanger tube blocking, instrumentation plugging or pipeline blocking, and internal corrosion; hence, the conditions suitable for their formation should be avoided whenever possible.

7.6.1.1 Methanol Injection

Methanol has a low freezing point. It can be injected at any temperature, and it is very effective as a hydrate inhibitor due to its ability to achieve high dew point suppression. Methanol is preferred for use at low-temperature conditions when there is separation equipment downstream, because glycol is harder to separate at low temperatures (due to its increased viscosity).

Higher injection rates of methanol are required since some is lost to the hydrocarbon gas phase (due to its higher volatility), and some is lost to the liquid hydrocarbon phase (due to its solubility in liquid hydrocarbons). It is only the dissolution of methanol in water that inhibits hydrate formation.

The efficiency of inhibition depends on the concentration of methanol injected. It can be injected intermittently or continuously.

7.6.1.1.1 Problems with Methanol Injection

Methanol can dissolve alcohol-based corrosion inhibitors injected into the process to prevent corrosion leading to unexpected corrosion problems.

Methanol can concentrate in the liquefied petroleum gas (LPG) stream. LPG is made largely of propane and mixed butanes. Methanol forms azeotropes with the propane and butane components in the LPG, and these cannot be separated by distillation.

7.6.1.1.2 Methanol Recovery

Recovering methanol from the process by distillation is not economical, so in most instances it is not recovered after use. In the gas plant methanol is not recovered.

7.6.1.2 Glycol Injection

Ethylene glycol (EG) is the most used of all the glycols in hydrate inhibition due to its low cost, lower viscosity, and lower solubility in liquid hydrocarbons. Losses are generally very small, so they do not need to be considered when calculating injection rates (see Figure 7.13).

7.6.1.2.1 Glycol Recovery and Regeneration

Glycol is recovered and regenerated. After injection into the process, the glycol–water solution and the liquid hydrocarbons can form an emulsion during agitation or expansion from a higher pressure to a lower pressure such as across a throttling valve. To recover the glycol, glycol and the condensed water mixture is separated from the gas stream in a separator. The recovered glycol–water mixture is then taken to the glycol regeneration unit where the water is stripped from the mixture, and the regenerated glycol can then be reused.

The regeneration process is designed to produce a glycol solution that must have a freezing point below the minimum temperature encountered in the system it is to be injected. The freezing point of a glycol solution in water is dependent on the weight percent of glycol in the solution. This is typically 75–80 wt.%.

Glycol injection also aids in dehydrating the gas.

As noted earlier, the viscosity of glycol increases as temperature decreases. The design of units containing chillers and refrigeration units where glycol is injected needs to consider this: If the temperature is too low, the rich glycol solution leaving the unit is very viscous, making downstream separation difficult.

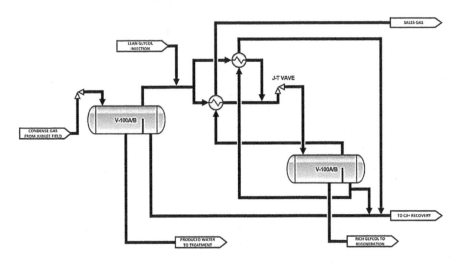

FIGURE 7.13   Glycol injection in the gas process plant.

7.6.2   GLYCOL DEHYDRATION

Avoid operational conditions that might cause the formation of hydrates by depressing the hydrate formation temperature using glycol dehydration. Refer to Section 3.3.3.4.2.1 above for further information on glycol dehydration.

7.6.2.1   Water Removal

If enough water can be removed from the produced fluids, hydrate formation will not occur. Dehydration is the common hydrate prevention technique applied to export pipelines.

For subsea production systems, subsea separation systems can reduce water flow in subsea flowlines. The advantage of applying subsea separation is not only hydrate control, but also increasing recovery of reserves and/or accelerating recovery by making the produced fluid stream lighter and easier to lift. These systems need to be combined with another hydrate prevention technique such as continuous injection of a thermodynamic inhibitor or low dosage hydrate inhibitor. The main risk associated with these subsea water separation systems is reliability (Yong, Qiang 2005).

7.6.3   LOW PRESSURE OPERATION

The pressure is maintained below the value necessary for hydrates to form. Decreasing the pressure below the pressure indicated on the hydrate formation curve for the given operating temperature.

7.6.4 Insulation

Insulation provides hydrate control by maintaining temperatures above hydrate formation conditions. Insulation extends the cooldown time before reaching hydrate formation temperatures.

When a pipeline carrying hydrocarbon fluid is shutdown for an extended period of time, the temperature in the system will eventually come to equilibrium with the surroundings. Depending on the nature of the hydrocarbon being transported, remedial actions may be necessary during the period of shutdown before the temperature reaches the minimum allowable value. It is thus of interest to be able to predict, with reasonable accuracy, the length of time the fluid will take to cool down to any particular temperature. To assure minimum cool down time, the simplest solution is to insulate the pipe. Since the flow condition will change during the field's life, insulation and/or active heating should be adapted to the different conditions.

Insulation is normally not applied to gas production systems, because the production fluid has low thermal mass and also will experience JT cooling. For gas systems, insulating is only applicable for high reservoir temperatures and/or short tieback lengths (Yong, Qiang 2005).

7.6.5 Active heating

Active heating includes electrical heating and hot fluid circulation heating in a bundle. In flowlines and risers, active heating must be applied with thermal insulation to minimize power requirements (Yong, Qiang 2005).

7.7 PREVENTION OF HYDRATE FORMATION DURING COMMISSIONING

7.7.1 Pipelines

For pipelines that normally transport dry gas, the most likely occasion at which hydrate formation may pose a problem is during commissioning. Prior to commissioning with hydrocarbon gas, the pipeline will have been filled with water in order to carry out a system hydrotest. Hence, between the hydrotest and gassing-up operations free water must be removed or treated to prevent the formation of hydrates. Methanol (MeOH) or glycol will normally be used to dewater equipment that carries gas (pipelines, jumpers, etc.) to prevent hydrate formation during the bulk dewatering phase of the commissioning process.

Numerous methods of drying or treating the pipeline available are:

- Methanol/glycol swabbing
- Vacuum drying
- Hot air drying
- Inert gas drying

For long pipelines, methanol swabbing offers the most economical solution. After a completed pipeline is filled with water and hydrostatically tested, conventional pigging runs remove the bulk of the water. Inevitably, there is a certain amount of water which passes the pigs and remains in the line. This residual water might form solid hydrates with the methane in natural gas under certain temperature and pressure conditions. Therefore, we use methanol or glycol swabbing, which involves passing a quantity of methanol down the pipeline between pigs.

This procedure leaves an aqueous film of methanol and water on the pipe wall that has a suitable concentration of methanol to inhibit hydrate formation when gassing-up. (Any residual water present dissolves in the methanol, and any fluid passing the pigs contains only a small proportion of water but a large proportion of methanol; this helps prevent any subsequent formation of hydrates.) From the volume of methanol used, the volume of liquid pigged out, and the percentage of methanol in this liquid, estimates can be made of the volume of water still in the main, how long it will take to become dry, and whether the methanol content of the aqueous film is high enough to prevent hydrate formation. Decisions can then be taken to either gas the line up or pig again with methanol. After commissioning the concentration of methanol in the aqueous film must be maintained until the main is dry. This can be achieved by either adding enough methanol to the gas or adding more methanol at intervals to the line.

Mono-ethylene glycol (MEG) is usually the fluid of choice for swabbing. MEG is typically chosen because it is easier to handle than methanol and can be regenerated.

The other drying procedures prevent the possibility of hydrate formation by completely vaporizing all free water in the pipeline.

7.7.2 SPOOL PIECES

The final connection to the pipeline system usually comprises a double block valve and bleed system. During the pre-commissioning operations, both of these block valves will usually be closed, leaving the spool piece between the valves full of water.

In order to prevent the possibility of hydrate formation during gassing-up, this water must be treated. The simplest solution is to fill the spool piece with either methanol or glycol to a suitable concentration that will inhibit hydrate formation at the pipeline operating conditions. Ideally, as much water as possible should be displaced by the methanol or glycol, as this will further minimize corrosion problems that may occur if acid gases dissolve in the free water.

7.8 REMOVING HYDRATES

The existence of hydrocarbon hydrates in a system may be detected by a change in the pressure profile within the system. Hydrates, like any other obstacle in a pipeline, can be detected by the consequences they create. The following may indicate the presence of hydrates:

- Reduced flow rates
- Reduced pressure

- Increased back pressure on a system
- Increased differential pressures
- Temperature drops

Once hydrocarbon hydrates have started to form, the build-up accelerates. Hence, the system pressure profile will continue to change, giving rise to potential capacity limitations.

The following methods can be used to remove hydrate plugs from a pipeline:

1. Depressurization: the most common way to remove a hydrate plug from a flow channel is by depressurization. Flow is stopped, and the line is slowly depressurized from both ends of the plug. At atmospheric pressure, the hydrate stability temperature is invariably less than that of the surroundings, so heat flows from the environment into the hydrate plug. The plug melts radially inward, detaching first at the pipe wall.

2. Another method of removal of hydrocarbon hydrates from a pipeline system involves running a batch of either methanol or glycol through the pipeline, driven by a pig. The size of the batch should be sufficient to depress the hydrate formation temperatures of the maximum anticipated hydrate deposition to below the pipeline temperature, as well as allowing for liquids that will be left on the pipe wall and not be pushed forward by the pig. Methanol or glycol injection is usually ineffective because of the necessity of having the inhibitor contact the hydrate-plug face. When hydrates form in a vertical portion of a channel, such as a riser or well string, it may be possible to inject glycol or place a heater at the plug face to promote the dissociation of hydrates.

3. Removal of the hydrates from raw natural gas pipelines, which may not have pigging facilities, is usually achieved by increasing the dosage rate of hydrate point depressant. For a short period, shock dosing, at a rate of approximately five times the normal injection rate, should be undertaken.

4. An alternative approach to removing hydrates is to operate the pipeline at different physical conditions. By lowering or increasing the operating temperature, the physical conditions suitable for hydrate formation may be avoided. This solution may be of limited use as the flexibility of the operating conditions, consistent with the required production rate, is unlikely to be high.

5. Coiled tubing represents the primary mechanical means of freeing the hydrate plug, but the maximum coiled-tubing distance is currently approximately 5 miles. Coiled tubing may be used to remove a substantial liquid hydrostatic head at the hydrate face to enable depressurization. Coiled tubing may also be used to inject methanol or glycol at the face of a hydrate plug when density is insufficient to drive the inhibitor to the plug face.

6. Troubleshoot the reason for the hydrate or ice plug formation and remedy the problem, if possible.

7.8.1 INJECTION OF HYDRATE POINT DEPRESSANT

Pipelines that are operated in the presence of free water at hydrate-forming conditions must be injected continuously with a hydrate point depressant or hydrate inhibitor. The most common hydrate inhibitors used are methanol, ethylene glycol, or diethylene glycol.

At pipeline conditions below −10°C, glycols are not often used due to their high viscosity. Above −10°C glycols may be preferred due to their lower injection rate requirement. The required injection rate of glycols is generally less than that for methanol due to the lower vaporization rate of the glycols into the gas. The minimum amount of hydrate inhibitor required can be calculated using Hammerschmidt's (1939) method in Equation (7.11) below (Maurice, Ken 2011):

$$\Delta T = \frac{KW}{100(MW) - (MW)(W)} \tag{7.11}$$

Where:

ΔT = the hydrate temperature depression from the equilibrium temperature at a given pressure (°F)

MW = molecular weight of the inhibitor (methanol = 32.0)

W = wt% of the inhibitor in the final water

K = constant, from Table 7.2 below

TABLE 7.2

Molecular weight and Hammerschmidt constant of various inhibitor

Inhibitor	Constants	
	MW	K
Methanol	32.04	2335
Ethanol	46.07	2335
Isopropanol	60.10	2335
Ethylene Glycol	62.07	2200
Propylene Glycol	76.10	3540
Diethylene Glycol	106.10	4370

Rearranging the Hammerschmidt equation (Equation 7.11) to find W:

$$W = \frac{100(MW)\Delta T}{(MW)\Delta T + K} \tag{7.12}$$

7.8.1.1   Determination of Total Inhibitor Required

$$\text{Total inhibitor required} = \left(\text{Inhibitor required freewater}\right)$$
$$+\left(\text{Inhibitor lost to vaporphase}\right) \qquad (7.13)$$
$$+\left(\text{Inhibitor soluble condensate}\right)$$

Where:

Inhibitor lost to vapor phase is determined from Figure 7.14. Methanol lost to
   the vapor phase
Inhibitor soluble in the condensate is approximately 0.5%

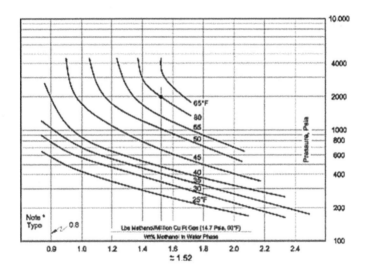

FIGURE 7.14   Ratio of methanol vapor composition to methanol liquid composition-example.

8 Leak and Break Detection

8.1 INTRODUCTION

Pipeline operating companies are required by codes/standards to make periodic pipeline balance measurements to check system integrity. Both installed devices and operational procedures must be in place to detect pipeline failures early.

Operations personnel must be diligent in the observation of pipeline and pipeline system components during field surveillance. Knowledge of normal operating conditions, such as system pressures, is integral to leak detection. Not all pipeline leaks are noticeable by operating conditions. Therefore, during daily rounds, the operator must observe line and lease conditions that may result in a failure.

Production volume discrepancies must be taken into account daily, since low production for no apparent reason may signify a pipeline leak or rupture.

In the event of a report of a problem or spill from the public or another outside party, the operator must immediately investigate.

If a pipeline leak or rupture is detected, the source of the released product must be isolated immediately. If there are multiple possibilities, isolate all possible sources and determine the correct source after the release is under control.

There is no de facto standard or accepted single method for leak detection, but the advance of simulating software and hardware brings new developments to the market each year and will need to be considered for each new pipeline.

Items to consider when investigating or specifying a leak detection system are:

- Sensitivity of the environment to a leak of the fluid
- Sensitivity required for the system to report a leak and the time taken to do it
- Number of sensor points available
- Availability of inspection information
- Ability to deal with transient conditions without casing spurious alarms
- Importance of the location of the leak (less for short pipelines, more for long ones)

8.2 LEAK DETECTION METHODS

Leak detection determines where a leak has occurred in liquid and gas pipeline systems. Methods of detection include hydrostatic testing (hydrotest), infrared thermography, and laser technology after pipeline erection and leak detection during service.

Causes of pipeline leakage can be divided into five main categories:

- Internal and external corrosion
- Third-party damage
- Operational error
- Natural hazards
- Mechanical failure

Leak detection for the pipeline may be achieved by the provision of equipment (e.g. Pipeline Integrity Monitoring and Leak Detection Systems (PIMS)) to undertake constant monitoring of pressure-sensing devices located at each end of the pipeline and at the intermediate isolating valve stations.

The output from these monitoring devices should be displayed in the main control room, thereby enabling the operators to identify abnormal or unexplained deviations in pressure and to shut in the pipeline section affected by actuating the intermediate/block isolating valves.

In addition to the above, in any PIMS/SCADA system, metering should be installed at each end of the pipeline. The signal from the meter at the receipt terminal should be transmitted to the main control room (usually the pumping end of the pipeline) for comparison with the outgoing meter signal. Any unexplained deviation from a predetermined threshold value should alert the operators as to a possible leak or malfunction.

The leak detection system may be classified as:

a. Internal-based leak detection system
b. External-based leak detection system

External-based leak detection system can detect the smallest leak with high accuracy. Internal-based leak detection discovers gas leakage based on measurement readings at some specific locations along the pipeline, e.g. computational pipeline monitoring (CPM).

The main aim of leak detection is to help pipeline operators to detect and localize leaks. The methods given in Table 8.1 may be used to detect leaks in pipelines.

Table 8.1 describes common techniques that should be considered for the detection of pipeline leaks.

TABLE 8.1

Summary of Some Leak Detection Techniques

Technique	Description
Soap solution bubble test	The pressurized unit to be tested is sprayed with a soap solution, and the operator is able to see the bubbles formed by gas escaping from where the leak is.
Water-immersion bubble test method	The water-immersion bubble test, also called "bubble testing" or "dunking", is a traditional and relatively primitive technique of leak detection. It consists of immersing a charged or pressurized part, usually with high-pressure dry air or nitrogen, in a water tank and watching for escaping bubbles.
Software-based leak detection system	Computational pipeline monitoring systems (also called software-based leak detection systems) use pipeline data to infer leaks on the pipeline and/or to alarm upon hydraulic anomalies that have the characteristics of a leak. These systems are in place to alert the pipeline controller so he/she can evaluate the cause and to shut down the pipeline as necessary and minimize the size of a spill.
H_2S detection	Permanent or portable tools are used to detect the presence of H_2S gas. Permanent monitors are used at surface facilities.
Right-of-way surveillance	Visual inspection by walking or aerial surveillance to look for any leakage along the pipeline. Any one of the following indications is a sign of a suspected natural gas pipeline leak: • Soil settlement; gas bubbling; and water, soil, or vegetation discoloration • Whistling or hissing sound • Distinctive, strong odor, often compared to rotten eggs • Dense fog, mist, or white cloud • Bubbling in water, ponds, or creeks • Dust or dirt blowing up from the ground • Discolored or dead vegetation above the pipeline right-of-way (ROW) Can be used in combination with infrared thermography and flame ionization surveys.
Production monitoring	Volume balancing (inlet and outlet flow) or pressure monitoring (pressure drop) is done to look for an indication of gas leakage. Changes in production volumes or pressure can indicate a pipeline failure. This is a more effective tool for finding large leaks and ruptures.
Detecting leaks through pressure changes	Leakage can cause a noticeable change in gas pressure. Therefore, sensors can be installed to detect changes in the pressure of the pipeline. Changes in pressure can trigger an alarm. The sensors required for this technique can be categorized as flow, pressure, and temperature.

(Continued)

TABLE 8.1 (CONTINUED)
Summary of Some Leak Detection Techniques

Technique	Description
Fiber-optic sensing technology	Leaks can cause sudden temperature changes in the soil surrounding a pipeline. Fiber-optic cables buried along pipelines can sense these temperature changes, as well as acoustic vibrations from a leaky pipe. Signals are then sent to the control room, and anything out of the ordinary triggers alarms.
PIMS/SCADA-based system	A leak detection system can be integrated into the SCADA system in the control room. SCADA systems can alert personnel in the control room whenever there is a leak, as well as facilitate record-keeping and trending before and after the event. Finding the leak's precise location facilitates quicker response and repairs.
Flame ionization survey	Electronic instrumentation is used to detect very low concentrations of gas leakage. Equipment is portable and very sensitive; the pipeline has to be displaced to a combustible gas. Equipment may be handheld, mounted on an All-Terrain Vehicle (ATV), or mounted to a helicopter.
Infrared thermography	Thermal imaging is used to detect temperature change on right-of-way due to escaping gas or produced water. A sufficient volume of escaping gas is needed to create an identifiable temperature difference. The process is normally completed using aerial techniques.
Odor detection	The gas odor is detected using trained animals and patented odorants. This method is able to detect pinhole leaks that may be otherwise non-detectable.
External leak detection equipment	External leak detection equipment can be installed on the pipeline. Detection equipment can monitor the dynamics of the flow for changes that would indicate a leak.
Intelligent pigging	Small leaks can produce ultrasonic signals which can be detected by a pig propelled forward by oil flow over several seconds, allowing several hundred samples. Very small leaks can be detected by this method. The disadvantage of this method is the frequent requirement of pigging.
Acoustic emission systems	The presence of a leak is manifested by an increased noise level. The sound generated by the leak can be used as a means of leak detection and its location.
Radioactive tracing	This technology is widely used to detect leakage of the process pipe and heat exchanger system pipe that carry liquid and gas in the oil and gas industry. To find a leak, the radioactive material is put into one end of the pipe. A radiation detector outside the pipe or above ground is used to track its progress through the pipe. The leak or blockage is discovered by finding where the amount of radiation gets decreased.

References

Ashworth, V., Booker, C. J. L.. 1986. *Cathodic Protection: Theory and Practice*. Birmingham: Institution of Corrosion Science and Technology.

ASME. 2010. *Gas Transmission and Distribution Piping Systems – B31.8*. New York: The American Society of Mechanical Engineers.

Baker Jr., Michael. 2008. *Pipeline Corrosion*. Integrity Management Program. Washington, DC: U.S. Department of Transportation.

BS 8010, Part 2. 2004. *Code of Practice for Pipelines - Part 2: Subsea Pipelines*. United Kingdom: BSI.

CAPP. 2018. *Mitigation of External Corrosion on Buried Carbon Steel Pipeline Systems*. Calgary, AB: Canadian Association of Petroleum Producers (CAPP).

CAPP. 2018. *Mitigation of Internal Corrosion in Carbon Steel Gas Pipeline System*. Calgary, AB: Canadian Association of Petroleum Producers (CAPP).

CAPP. 2018. *Mitigation of Internal Corrosion in Carbon Steel Oil Effluent Pipeline Systems*. Calgary, AB: Canadian Association of Petroleum Producers (CAPP).

Cato, Sky. 2019. "In-Situ Coating Process Using Pigs, that Can Internally Coat Pipelines and Carry Out Rehabilitation Works." *World Pipelines: Coating & Corrosion* (Palladian Publications Ltd) 7–11.

Chalke, P., Hooper, J. 1994. *Addendum to Corrosion protection Guidelines*. Middlesex: JP Kenny Ltd (unpublished).

Cheng, Frank, Qian, Shan. 2017. *Accelerated Corrosion of Pipeline Steel and Reduced Cathodic Protection Effectiveness Under Direct Current Interference*. Vol. 148, in *Construction and Building Materials*, by Shan Qian, Frank Cheng, 675–685. ScienceDirect. Accessed January 8, 2020. https://advanceseng.com/accelerated-corr osion-pipeline-steel-reduced-cathodic-protection-effectiveness/.

Dairyland. 2020. *Dairyland - Solid State Decouplers*. Accessed January 17, 2020. https://ww w.dairyland.com/knowledge-base-article/33-getting-started/106-what-are-solid-sta te-decouplers-and-how-are-they-used.

Davis, Gerald. 1994. "Classic Methods of Corrosion Control - Summarized." http://www. dmme-engineering.com/. Accessed May 29, 2020. https://www.experts.com/articles/cl assic-methods-corrosion-control-by-gerald-davis.

Dean, Hale. 1984. "Special Report-Slick Way to Increase Capacity." *Pipeline and Gas Journal* 17–19.

Gas Processors Suppliers Association. 1980, 2014. *S. I. Engineering Data Book*. Tulsa, OK: Gas Processors Suppliers Association (GPSA).

Ginzel, R. K., Kanters, W. A.. 2002. "Pipeline Corrosion and Cracking and the Associated Calibration Considerations for Same Side Sizing Applications." *NDT.net* 7. Accessed February 21, 2020. https://www.ndt.net/article/v07n07/ginzel_r/ginzel_r.htm.

GSA. 2012. "Natural gas pipeline safety (construction, operation and maintenance) regula-tions – L. I. 2189." 85. Ghana: Ghana Standards Authority (GSA).

Hilti. 2015. *Corrosion Handbook*. Boston, MA: Addison-Wesley Publishing.

HRR. 1992. *Onshore Pipeline Design Base Manual*. Design guideline. JP Kenny, Unpublished.

IGEM. 2008. *IGEM/TD/1 Edition 5 - Steel Pipelines for High Pressure Gas Transmission*. London: Institute of Gas Engineers.

IPLOCA. 2014. "Onshore Pipeline Survey Requirements by Project Phase." In *Onshore Pipelines the Road to Success*, by IPLOCA, 97. IPLOCA.

Javaherdashti, Reza, Nwaoha, Chikezie, Tan, Henry. 2013. *Corrosion and Materials in the Oil and Gas Industries.* Boca Raton, FL: CRC Press.

Jorda, R. M.. 1966. "Paraffin Deposition and Prevention in Oil Wells." *Journal of Petroleum Technology* (Society of Petroleum Engineers) 18 (12). Accessed 2018. doi: 10.2118/1598-PA.

Kadir, Ali. 2002. *Distribution, Transmission System and Design.* Lecture notes. Salford: School of Computing Science & Engineering - University of Salford.

Klohn, C. H. 1959. *Pipeline Flow Test - Evaluate Cleaning and Internal Coating.* Gas.

Kut, S. 1975. "Internal and External Coating of Pipelines." *First International Conference on Internal and External Protection of Pipes.* University of Durham: BHRA Fluid Engineering, 50–51.

Langill, Thomas J. 2006. *Corrosion Protection.* Course notes, Iowa City, IA: University of Iowa.

Lawrie, Dan. 2018. *Dan Lawrie Insurance Brokers.* July 23. Accessed March 28, 2020. http://danlawrie.com/mitigating-construction-defects-with-a-quality-control-program/.

Lidiak, P. 2010. "Hazardous Liquids Pipeline Industry Perspective on Excavation Damage." *Pipeline Safety Trust Conference.* New Orleans, LA. 4.

Lin, Tian Ran, Guo, Boyun, Song, Shanhong, Ghalambor, Ali, Chacko, Jacob. 2005. *Offshore Pipelines.* Oxford: Elsevier.

Mather, John, Blackmore, Chris, Petrie, Andrew, Treves, Charlotte. 2001. "An Assessment of Measures in Use for Gas Pipelines to Mitigate Against Damage Caused by Third Party Activity." Accessed September 10, 2018. http://www.hse.gov.uk/research/crr_pdf/2001/crr01372.pdf.

Mohammed, Reyadh M. M., Suliman, Saad M. A. 2019. "Delay in Pipeline Construction Projects in the Oil and Gas Industry: Part 1 (Risk Mapping of Delay Factors)." *International Journal of Construction Engineering and Management,* 24–35.

Muhlbauer, W. Kent. 2004. *Pipeline Risk Management Manual: Ideas, Techniques, and Resources.* Third edition. USA: Elsevier.

Nasr, G. G., Connor, N. E., Burby, M. L. 2003. *Basic Units of Measurement, Gas Supply & Combustion.* Lecture notes. Salford: School of Computing Science & Engineering - University of Salford.

Natarajan, K. A. 2012. "Advances in Corrosion Engineering." Lecture notes. Bangalore: NPTEL Web Course.

Norsworthy, Richard. 1996. "High Temperature Pipeline Coatings Using Polypropylene Over Fusion Bonded Epoxy." *International Pipeline Conference — Volume 1.* Calgary, AB: American Society of Mechanical Engineers (ASME), 253–260.

NWE. 2016. "Planting Trees Near Gas Pipelines." *North Western Energy (NWE).* Accessed October 2, 2018. https://www.northwesternenergy.com/safety/community-safety/planting-trees-near-gas-pipelines.

Nyman, Douglas J., Lee, Edward Mark, Audibert, Jean M. E. 2008. "Mitigating Geohazards for International Pipeline Projects: Challenges and Lessons Learned." *Proceedings of IPC 2008: 7th International Pipeline Conference.* Calgary, AB: IPC, 1–11.

OECD. 1997. "Report of the OECD Workshop on Pipelines (Prevention of, Preparedness for, and Response to Releases of Hazardous Substances)." *Organization for Economic Co-operation and Development (OECD).* Paris: OECD, 17–18. Accessed November 2019. http://www.oecd.org/officialdocuments/publicdisplaydocumentpdf/?doclanguage=en&cote=ocde/gd(97)180.

Okyere, Mavis Sika. 2015. *Pipeline Integrity Management Manual.* Accra: Ghana National Gas Company (GNGC).

O'Malley, Cindy. 2018. *High-Temperature Service (Heat-Resistant) Coatings: Industries, Coating Types, Testing Protocols and Consequences of Testing Inconsistencies.* March 16. Accessed September 27, 2018. https://ktauniversity.com/high-temperature-coatings/.

Overpipe. 2018. "Third Party Interference (TPI) on Buried Networks." *IPLOCA* 2–18.

Palmer-Jones, Roland, Paisley, Dominic. 2000. "Repairing Internal Corrosion Defects in Pipelines - A Case Study." *4th International Pipeline Rehabilitation and Maintenance Conference*. Prague: Penspen, 6–14.

PeaceFM. 2019. *Another Sabotage? Gas Pipelines at Tema Torched by Suspected Arsonists*. Accra, April 8. Accessed April 10, 2019. http://www.peacefmonline.com/pages/local/news/201904/379704.php.

Posakony, G. J. 1993. "Integrity Assurance of Natural Gas Transmission Pipelines." *ASME - Applied Mechanics Review (American Society of Mechanical Engineers (ASME))*, 146–150.

Safamirzaei, Mani, Namvaran, P. T.. 2015. "Predict Gas Hydrate Formation Temperature with a Simple Correlation." *Gas Processing & LNG*. July. Accessed May 13, 2020. http://www.gasprocessingnews.com/features/201508/predict-gas-hydrate-formation-temperature-with-a-simple-correlation.aspx.

Schmitt, G., Bakalli, M.. 2008. "Advanced Models for Erosion Corrosion and Its Mitigation." *Wiley*, 181–192.

Singh, Ramesh. 2014. "Stress Corrosion Cracking in Steels." In *Pipeline Integrity Handbook: Risk Management and Evaluation*, by Ramesh Singh, 95–102. Houston, TX: Gulf Professional Publishing.

Sloan, Riachard N. 2001. "Pipeline Coatings." In *Peabody's Control of Pipeline Corrosion*, by A. W. Peabody, edited by Ronald L. Bianchetti, 7–21. Houston, TX: NACE International.

Stancliffe, J. 2017. "Third Party Damage to Major Accident Hazard Pipelines." September 6. Accessed August 19, 2019. http://www.hse.gov.uk/pipelines/ukopa.htm.

Stewart, Maurice, Arnold, Ken. 2011. *Gas Dehydration Field Manual*. Houston, TX: Gulf Professional Publishing - Elsevier.

Singh, G., Samdal, O. 1987. "Economics Criteria for Internal Coating of Pipelines." *International Conference on the Internal and External Protection of Pipelines*. University of Durham: BHRA. Paper Europe 1.

Taiwo, Idowu Adeyinka. 2013. *The Effect of Bitumen Coatings on the Corrosion of Low Carbon Steels (API 5L X65)*. MSc Theisis, Abuja: African University of Science and Technology.

Thomas, Laura, Burden, Tim. 2010. *Heavy Oil Drag Reducing Agent (DRA): Increasing Pipeline Deliveries of Heavy Crude Oil*. Houston, TX: ConocoPhillips Speciality Products Inc.

Tong, Shan. 2015. *Cathodic Protection*. Training document. Ghana: Sinopec.

Turkiewicz, Anna, Brzeszcz, Joanna, Kapusta Piotr. 2013. *The Application of Biocides in the Oil and Gas Industry*. Krakow: Oil & Gas Institute.

von Baeckmann, W., Schwenk, W., Prinz, W. 1997. *Handbook of Cathodic Corrosion Protection*. Houston, TX: Elsevier Science - Gulf Professional Publishing.

Yong, B., Qiang, B. 2005. *Subsea Pipeline and Risers*. London: Elsevier.

Zavenir, Daubert. 2018. *zavenir-blog*. June 1. Accessed September 22, 2018. http://www.zavenir.com/blog/types-of-corrosion-inhibitors/.

Index

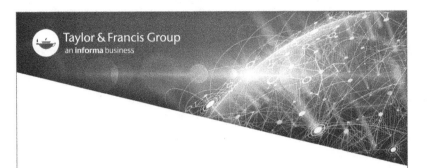